The Great Migration

Jonathan Scott

The Great Migration

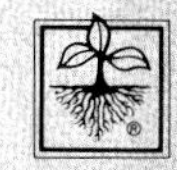

Rodale Press, Emmaus, Pennsylvania

First published in Great Britain 1988 by Elm Tree Books

Book Design by Trevor Vincent
Maps by Don Macpherson

Published in 1989 in the United States of America by
Rodale Press, Inc.
33 E. Minor St.
Emmaus, PA 18098

If you have any questions or comments concerning this book, please write:

Rodale Press
Book Reader Service
33 East Minor Street
Emmaus, PA 18098

ISBN 0-87857-866-8 hardcover

2 4 6 8 10 9 7 5 3 hardcover

Typeset in Linotron Palatino by Centracet
Printed and bound in West Germany
by Mohndruck, Gütersloh

Distributed in the book trade by St. Martin's Press

Contents

For the animals shall not be measured by man. In a world older and more complete than ours, they move finished and complete, gifted with extensions of the senses we have lost or never attained, living by voices we shall never hear. They are not brethren, they are not underlings; they are other nations caught with ourselves in the net of life and time, fellow prisoners of the splendour and travail of the earth.

HENRY BESTON – THE OUTERMOST HOUSE

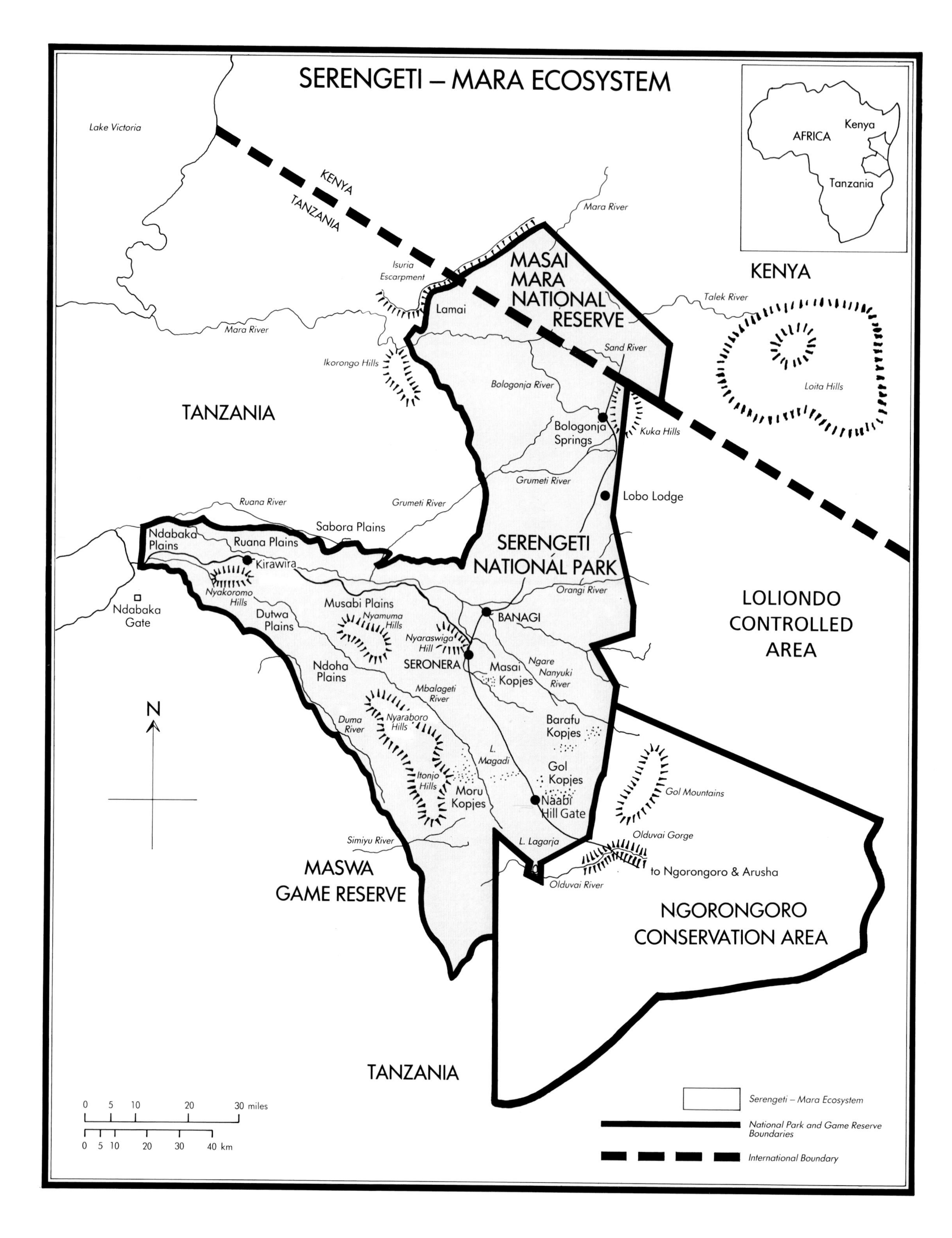

SERENGETI – MARA ECOSYSTEM
AFRICA
Kenya
Tanzania
Lake Victoria
KENYA
TANZANIA
Mara River
Isuria Escarpment
MASAI MARA NATIONAL RESERVE
KENYA
Talek River
Lamai
Mara River
Sand River
Ikorongo Hills
Bologonja River
Loita Hills
TANZANIA
Bologonja Springs
Kuka Hills
Grumeti River
Lobo Lodge
Ruana River
Grumeti River
Sabora Plains
Ndabaka Plains
Ruana Plains
SERENGETI NATIONAL PARK
Kirawira
Orangi River
LOLIONDO CONTROLLED AREA
Nyakoromo Hills
Ndabaka Gate
Musabi Plains
Dutwa Plains
Nyamuma Hills
BANAGI
Nyaraswiga Hill
SERONERA
Ndoha Plains
Masai Kopjes
Ngare Nanyuki River
Mbalageti River
N
Duma River
Nyaraboro Hills
Barafu Kopjes
L. Magadi
Gol Kopjes
Itonjo Hills
Moru Kopjes
Gol Mountains
Naabi Hill Gate
Simiyu River
L. Lagarja
Olduvai Gorge
to Ngorongoro & Arusha
Olduvai River
MASWA GAME RESERVE
NGORONGORO CONSERVATION AREA
TANZANIA
0 5 10 20 30 miles
0 5 10 20 30 40 km
Serengeti – Mara Ecosystem
National Park and Game Reserve Boundaries
International Boundary

Introduction

Men are easily inspired by human ideas, but they forget them again just as quickly. Only Nature is eternal, unless we senselessly destroy it. In fifty years' time nobody will be interested in the results of the conferences which fill today's headlines.

But when, fifty years from now, a lion walks into the red dawn and roars resoundingly, it will mean something to people and quicken their hearts whether they are bolshevists or democrats, or whether they speak English, German, Russian or Swahili. They will stand in quiet awe as, for the first time in their lives, they watch twenty thousand zebra wander across the endless plains.

BERNHARD AND MICHAEL GRZIMEK – SERENGETI SHALL NOT DIE

There is a place in northern Tanzania which the Masai call *siringet*. It means simply a wide open space. To the outside world and the thousands of tourists who visit *siringet* each year it is better known as Serengeti; the last place on earth where it is still possible to recapture a sense of our past, recalling the time when all of Africa's plains teemed with wildlife and man was still an integral part of the natural world.

There are few who are not profoundly moved by a visit to the Serengeti. For some it is a crossroads in their lives, an emotional rebirth, a lightening of the spirit at the sight of so many animals, such space.

I shall never forget my first view of those plains. Standing on Naabi Hill in the southern Serengeti, one morning in March, I looked out across a land that stretches uninterrupted as far as the eye can see. A view so immense that it defies description, offering no point of reference, no way of discerning scale or distance – as if God had steam-rolled this particular part of Africa into an enormous plain, then sown its soils of wind-blown ash with a rich blanket of short green grass.

By October – the height of the dry season – those same plains look desolate, providing barely enough food for ostriches and Grant's gazelles, dry country species that can live independent of drinking water. Yet I knew if I returned in January I would find the land awash in a sea of animals. That is the time when the Serengeti's short grass plains play host to the great spectacle known as the migration: hundreds of thousands of wildebeest and zebra spread evenly across the land. I could already see the vast numbers, hear the incessant cacophony of the great herds, smell the dust churned up by more than a million pairs of hooves. And moving among them – giving the whole scene an exquisite air of tension – the predators that have always enthralled me: lions and hyaenas, cheetahs and jackals, and – scarcest of all – the elusive wild dogs.

The Serengeti is part of Masailand, and red-cloaked warriors certainly trekked across these plains long before the first European explorer, Dr Oscar Baumann, set foot in the area in 1892. But it was not until the late 1950s and the arrival of another pioneering German, Dr Bernhard Grzimek, and his son Michael that the Serengeti became known throughout the world. Together they undertook the first aerial census of the wildebeest. Tragically Michael plunged to his death in the Grzimeks' zebra-striped aircraft after a mid-air collision with a griffon vulture. But their work was virtually completed, and in 1959 the publication of *Serengeti Shall Not Die* became a landmark in the fight to save Africa's wild animals.

Nearly thirty years after Michael Grzimek's death, how was I going to tell the story of the Serengeti, to convey the majesty of this last place? One can look at lions and leopards, view them in isolation and marvel at their beauty. But they are only one small part of the story. The odd-shaped wildebeest are the key. They *are* the Serengeti, inadvertently conditioning the lives of the other animals with whom they share the land. Its existence as we know it depends on them.

In early 1983, I was fortunate in being able to make a safari to the Serengeti. It was then that I first saw the wildebeest massed on the plains around Naabi Hill. Despite years spent living in the Masai Mara, just across the border in Kenya, I had never seen anything to rival that sight. From that moment I was determined to try and document the story of the great migration. By following the wildebeest on their journey, I would be compelled to see everything – the complex web of life, the interrelation of it all.

The history of the migration is as ancient as man. Fossil evidence from Olduvai Gorge, the most famous archaeological site in the world, indicates that *Connochaetes taurinus*, the modern wildebeest, seasonally grazed the Serengeti plains more than a million years ago. Over countless eons the wildebeest population has ebbed and flowed, shrinking in the face of disease and droughts, then expanding again in years of good rainfall, responding all the time to the ever-changing mosaic of grass and woodland.

Zebra migrate too, though their population is but a fraction of the wildebeest's: 200,000 compared to 1.3 million. If there is an underlying sense of orderliness about the movements of the zebra families – whether crossing a river or making their way through the long grass – there is an air of utter madness about the behaviour of the great herds of wildebeest. You would think they were possessed of demons, or had an urgent appointment with some hidden ruler.

At present the wildebeest move through a region that extends from the Crater Highlands in the east almost to the shores of Lake Victoria in the west, and from the Eyasi Escarpment in the south, all the way north to the Mara country in Kenya – an area of approximately 30,000 square kilometres. Their annual journey through this vast region – which may involve an individual in a trek of up to 3,000 kilometres – has long been thought to retrace age old migration routes, implying a degree of predictability about their behaviour. Yet wherever I read or inquired, I found different answers to my questions: which month is best to see the vast armies of animals massed on the plains; when and where the wildebeest calve; when to witness the dramatic exodus of wildebeest and zebra from plains to woodlands, and where best to stand in awe to see thousands of wildebeest storm across the Mara River.

The odd-shaped wildebeest are the key . . .

 There is an underlying sense of orderliness about the movements of the zebra families . . .

I soon discovered that you could spend a lifetime in Serengeti/Mara waiting for a typical migration. The finer details of the herds' movements are always different. It is a dynamic process which defies predictions: no two years are ever quite the same. Wildebeest are nomadic, responding in a highly opportunistic manner to the vagaries of rainfall and the subsequent availability of food.

Some years are wetter than others: it may pour for days on end, from one month to the next throughout the rainy season. In another year, perhaps even the next, the land can become a dustbowl. And it is this which determines the timing of the wildebeest's movements. People may assure you that the wildebeest will be massed on the southern plains of the Serengeti from December to May. But if the rains are insufficient to keep the short grass growing and the shallow waterholes dry up, then the wildebeest will retreat to the woodlands. With the most informed advice you could still miss the peak of the calving out on the plains; the wildebeest might vanish into the woodlands beyond Seronera weeks earlier or later than predicted; and you could camp on the banks of the Mara River throughout September yet still fail to witness a spectacular river crossing.

No account of the Serengeti would be complete without including Kenya's Masai Mara National Reserve, a 1,510 square kilometre sanctuary on the Kenya/ Tanzania border. Together – for they are but one place – Serengeti and Mara comprise the greatest game viewing area in the world, though they represent only half the area defined by the wanderings of the migratory wildebeest.

Despite their protected status, neither area has escaped the poachers. The five hundred black rhino who lived in the Serengeti not so long ago have all but vanished, and elephant numbers have been drastically reduced.

It might be tempting to feel despondent, to think that yet another 'last Eden' was crumbling under the heavy hand of man. The over-riding impression created by the Grzimeks' Oscar-winning film of *Serengeti Shall Not Die* was of the magnitude of it all: a land still swarming with game; the largest such concentration to be found anywhere in the world. I well remember feeling awed by the sheer number of animals that flitted across the screen. Yet the Grzimeks, it transpires, made their epic film around the time of the last great outbreak of rinderpest, the killer plague that has at varying intervals decimated the wild herds of wildebeest and buffalo. They were present when numbers were a fraction of current populations.

So today, as people increasingly voice their concern over what remains of Africa's wildlife heritage, the Serengeti and Mara harbour more of most animals than at any time within living memory. For the moment, the land of the great migration has survived.

The lioness crouched among the smooth-surfaced rocks of the granite outcrop . . .

1 *The Birth of the Migration*

The survival of our wildlife is a matter of grave concern to all of us in Africa. These wild creatures amid the wild places they inhabit are not only important as a source of wonder and inspiration but are an integral part of our natural resources and of our future livelihood and well-being.

MWALIMU JULIUS K. NYERERE, FIRST PRESIDENT OF TANZANIA

Two tawny ears were all that were visible of the lioness crouched among the smooth-surfaced rocks of the granite outcrop. She gazed into the distance, sharp eyes narrowed to golden slits, her attention riveted on the column of black specks emerging out of the dawn on the eastern horizon. Every so often she shook her grizzled head, temporarily dislodging the horde of bothersome flies that clung to the edges of her ragged ears. Even at that great distance she recognised the approaching animals as wildebeest. They were one of her most favoured prey and, old as she was, she was still quite capable of killing for herself.

The rocky island which hid the watching lioness rose like some medieval fortress among the vastness of the Serengeti plain. Besides lions, such outcrops or kopjes harbour an extraordinary variety of creatures, some of which spend their entire lives cloistered within these secret environs. No single group of kopjes is quite the same as another. Each, in its own way, acts like a stone ark anchored in a sea of grass.

Distinctive clusters of kopjes bear names bestowed on them long ago by nomadic tribesmen and adventurers of an earlier era: Masai pastoralists, trophy hunters, game wardens. To the west are the Moru Kopjes, a sprawling complex of rock shrouded by thick vegetation where plains and woodlands merge. Here each year a pair of Verreaux's eagles build their huge stick nest on a pedestal of rock, soaring overhead on long black wings in daily search of an unwary hyrax. To the east stand the low flung Gols, gaunt, lonely shapes set among the short grass plains, where cheetahs often hunt. Near the centre of the park – within sight of Seronera Lodge – the Masai Kopjes attract bus-loads of tourists who gather expectantly each morning and evening in the hope of catching a glimpse of the leopard and her two boisterous cubs which have temporarily taken up residence there. And, close to the main road between Naabi Hill and Seronera, are the Simba Kopjes, jutting from the earth thirty metres high in one place, and named for the lions that are so often found there, keeping watch over the plains from some cool place of concealment.

The wildebeest moved steadily closer, breaking into a gallop as they shied away from a small pack of muddy-coated hyaenas, whose thick hairy necks were matted red with the blood of last night's kill. But the predators hardly seemed to notice the passing herd, content for the moment to slumber full-bellied in their favourite wallow. The wildebeest had spent the night feeding on the higher ground to the south. Now they were headed towards a pool of sparkling blue water where zebra had already gathered to drink in the shadow of the Simba

Kopjes, and where the solitary lioness lay waiting.

The herd paused, the perfect symmetry of the column broken as the stragglers fanned out where others had stopped to feed. I could see them more clearly now, strangely-fashioned creatures with black manes and long white beards, their sleek bodies coloured varying shades of grey by the morning light. There was nothing to distinguish one from another. Only the yearlings stood out as somewhat smaller, with horns less heavily curved. All the animals moved at the same pace, paused to feed in unison, then galloped on again, obedient to some hidden command that ensured the integrity of the herd.

Once more, the wildebeest stopped to feed. The lioness tensed as they edged closer. Had she already noticed something, a slight defect, a diseased or aged animal, a vulnerability hidden from the human eye that would trigger an attack?

Then I saw what the lioness was looking at. Encircled by a thicket of legs, one of the cows had lain down. In the eight and a half months she had carried her calf she had survived the ever-present threat of predators, poachers and river crossings. Now, far from the woodlands where the resident leopards and lions might most easily ambush her, she was ready to give birth.

The cow's black legs glistened. Her waters had already broken, and a pair of tiny pale hooves were clearly visible beneath her tail. In an instant the lioness was on her feet. She descended quickly from the rocks and crept around the back of the kopje in a wide arc. When she reappeared she paused, assessing the situation before slinking forward again. Once more she hesitated.

Heads bowed to the ground, the wildebeest remained oblivious to the danger. Not one animal looked up as the lioness crept into the open, taut as a coiled spring, the thick muscles of legs and shoulders starkly profiled beneath her pale fur. But just as she gathered herself for the low run that would take her to within striking distance of the cow, a zebra stallion noticed the stealthy movements of the advancing predator. He snorted loudly and in a trice the cow was on her feet, responding instantly to a familiar call of alarm. The lioness froze in mid-stride, staring unblinkingly at the wildebeest. All eyes focused in her direction. But when she started forward again the herd turned as one and fled.

Driven on by the hunger of three days without even the scraps stolen from a hyaena's kill, the old lioness pursued the wildebeest, straining every muscle in her efforts to narrow the gap. But despite the imminent birth of her calf, the cow had the stamina to out-distance the lioness and leave her panting in a cloud of choking dust. She had been saved by her ability to interrupt labour when danger threatened.

An hour later the cow lay down again. This time there were no lions or hyaenas to threaten the safety of her calf. The contractions were coming faster, forcing her to lie flat on her side and stretch her thin legs out stiffly from her swollen body. She heaved and strained, holding her head and neck out taut, briefly exposing her white teeth in a grimace of effort – almost as if she were departing from life rather than giving birth to it.

The calf rose from the ground on shaky limbs, precariously swaying from side to side, back legs splayed wide . . .

Gradually the calf's dark knobbly head appeared – too late now for the cow to move on if a lioness found her – then its slender body, sheathed in the translucent foetal membranes. At the moment of birth, the cow stood up and the calf slumped limply on to the dry ground, severing the cord. Almost before the mother had lowered her head to lick and clean the calf, it jerked its spindly legs and sat up, its long droopy ears giving it a dog-like appearance. These first few minutes of life are vital in the mother/calf relationship, precious minutes in which the cow learns that this calf is hers, imprinting its sight, smell and taste on her senses. With her broad-lipped muzzle, the cow quickly nibbled away the enveloping membranes, stimulating the calf to try and stand.

At first the other members of the herd ignored the cow, except for one inquisitive animal that briefly sniffed the damp earth where the cow had been lying. Now as the calf struggled to rise, half a dozen heavily pregnant females crowded around, fascinated by the sight and smell of this first arrival. Could it be that this in some way helps to induce labour in other cows? One exuberant beast performed a frenzied dance of excitement, bucking and tossing its ungainly head. Another, this one a yearling, gently butted the calf on its quivering rump just as it seemed about to stand for the first time.

 Thousands of wildebeest insist on crossing Lake Lagarja each year

Twice more, the calf rose from the ground on shaky limbs, precariously swaying from side to side, back legs splayed wide. And each time, as it tried to step forward, it pitched unceremoniously on to its chin. But the ancestral urge to be on its feet and running with the herd forced it up once more. This time it remained standing, just five minutes old. Moments later it discovered its mother's teats and drew its first milk, another essential part of the process which binds calf to cow.

Vultures and eagles soared overhead, but there was nothing for them here. It would be another hour or so before the cow expelled the afterbirth. Then the scavenging birds would plummet from the sky, to squabble over the nutritious placenta. And in their wake the jackals and hyaenas would come running. Nothing would be wasted. But by then the calf would be strong enough to keep up with the herd.

During the next few days I watched many wildebeest giving birth and saw countless others with foetal sacs or legs protruding, betraying their condition. Each time the herd paused to feed, these expectant mothers lay down. But as soon as the other wildebeest drifted away, isolating the cows, they got to their feet and hurried to catch up, anxious to immerse themselves once more within the security of the herd. Those with newborn calves sought the company of others, forming small scattered groups within the larger aggregations, and thereby decreasing the chances of their calf being singled out by a hungry predator. Soon the herds seethed with tens of thousands of tiny buff-coloured, black-faced calves and you could be forgiven for thinking the wildebeest a beautiful creature.

Many people ask where the migration begins. Surely there must be a start and a finish? But there is no real beginning or end to a wildebeest's journey. Its life is an endless pilgrimage, a constant search for food and water. The only beginning is here, at the moment of birth.

That moment is so carefully synchronised that eighty per cent of pregnant cows drop their calves within a few weeks of each other – usually between late January and mid-March. Almost all cows over three years old produce a calf. In recent years the peak has been less pronounced – with such vast numbers involved, it is perhaps more difficult to synchronise the rut so perfectly.

The marked seasonal peak in births ensures that young calves are unavailable to predators for much of the year – they are all vulnerable at the same time, and in the same part of their range: the predators are glutted by the sheer volume of calves. And therein lies the individual's best chance of survival. Those born 'out of season' are invariably taken by predators. A calf's protection lies in its ability to gain co-ordination faster than any other ungulate, in being constantly on the move, and in seeking the anonymity of the herd. But many still die before their first year's journey has even begun, victims of under-nutrition, disease and predation.

Among such enormous numbers of animals, it is all too easy for a calf to become separated from its mother. Predators and cars sometimes cause the herds to panic, scattering wildebeest in every direction. Sometimes a mother and calf become separated when they ford a river, or as they cross Lake Lagarja, on the southern boundary of the park – which thousands of wildebeest insist on doing each year, despite the fact that they could easily walk around it.

Calves bleat when they are lost, and it is thought that a cow – given time – learns to recognise the sound of her calf as well as its smell. Orphans wander for days through the herds, desperately trying to re-find their mothers. But no cow wildebeest will accept the calf of another, even if she has recently lost her own calf and still has abundant milk with which to suckle an orphan. Consequently, these calves pass from cow to cow, each of whom lowers her head to smell them before nudging them on their way again. Some calves are undeterred, following for several kilometres in the footsteps of a cow and her calf. Even bulls receive tentative advances from lost calves, who soon find themselves rebuked for their error with a rough sweep of the male's horns. Just occasionally, the seemingly impossible happens and a calf finds itself reunited with its mother. But many never do.

Sometimes two orphans will race across the plains towards each other, finding brief comfort in the presence of a fellow creature. But they do not consort for long before rushing off again in search of something more closely resembling the size of their mother – even if not the shape. Too young to have benefited from the cumulative wisdom of the herd, calves blunder towards hyaenas already fat

Hyaenas are capable of killing adult zebra . . .

Just occasionally a calf finds itself reunited with its mother . . .

from days of feasting, or trot blithely into the fatal embrace of lions. One huge male, tawny mane raised in a conspicuous mantle around his head, tried as best he could to flatten himself against the ground as a calf raced towards him. He hardly had to move – a short rush and the calf lay dead, just metres from the half-eaten carcass of an earlier victim. Others followed my car, matching their pace to mine, galloping alongside for a while before turning away to stand alone in the vastness of the Serengeti.

Spotted hyaenas are by far the most numerous of all the Serengeti's larger predators. More than three thousand of these formidable creatures spend the wet season on the plains, carefully tracking the movements of the wildebeest as they wander across the grasslands.

Hyaenas hunt mainly under cover of darkness, though they are great opportunists, responding quickly, day or night, to the presence of a vulnerable prey animal. They do not require long grass or thick bush for cover. Their methods are those of a courser, relying on speed and dogged endurance in the chase. When hunting in groups, hyaenas are capable of killing adult wildebeest and zebra. Even a single hyaena sometimes pulls down a full-grown wildebeest. If it does, other hyaenas soon arrive and together they tear their victim apart.

. . . stimulating the calf to try and stand . . .

When the wildebeest are giving birth on the plains, the hyaenas take a heavy toll of the new arrivals. It is noticeable that when danger threatens, a cow will move away with her calf shielded on the far side of her body. The calf stays close to its mother's shoulder, and can be difficult to identify behind the shroud of the cow's long beard. It is a common sight early in the morning to see a hyaena barrelling across the plains in pursuit of a wildebeest and her calf. For the first few days of life a calf is somewhat slower than its mother, and it is these youngsters that the hyaenas prefer to hunt. But within a week, calves can gallop at speeds of fifty kilometres an hour. This is their best defence – to try and outrun the predators.

A cow wildebeest is quick to defend her calf against an attack by hyaenas and wild dogs, though she receives no support from other herd members. At the height of a chase, she may fall back a few paces, trying to block the path of a pursuing hyaena, something she would never dare try against a lioness. But if there are two or more hyaenas hunting together it often proves impossible for the mother to keep them all at bay – whilst she charges and hooks at one hyaena, another dashes in and boldly snatches up her calf. One bite in the neck from those powerful jaws is usually sufficient to end the struggle. As soon as the calf stops bleating, its mother's aggressive instincts subside, though I often saw such females desert the herd to wander back and forth across the plains for hours in search of their calf.

A more subtle form of protection against predators, and in particular against hyaenas, is the time of day when most calves are born. Some are born at night, while shrouded in darkness. But the majority of births I have witnessed occur around the middle of the morning, even at mid-day. By this time most hyaenas are safely out of sight, seeking the shade provided by shallow erosion terraces or lying belly-deep in mud wallows, and the lions are sprawled among the cool granite rocks of the nearest kopje. Calves born in mid-morning have the rest of the day to gain strength before the predators rouse themselves to hunt again.

One sees death everywhere during this time of rebirth for the wildebeest nation. Eventually it appears commonplace, no longer shocking. All the larger predators turn their attention to the wildebeest calves. Lions, hyaenas, cheetahs, wild dogs – all killing. Many visitors are horrified at the thought of so much butchery, while still harbouring hopes of seeing the predators hunting. Our own predatory nature is rooted deep within us.

Perhaps our aversion to the killing of one animal by another is heightened by the ease with which we now obtain a meal. We are no longer required to hunt for food. It can be purchased in a tin or at a restaurant. The distasteful process of killing to eat is removed for most of us. But surely no justification is required of the predators. They kill to eat, that is all.

 The Serengeti's animal-speckled plains . . .

2 *Serengeti – the Beginning*

Would the animals be able to go on living here? Were there enough plains, mountains, river valleys and bush areas to maintain the last giant herds still in existence? We had already noticed that large herds of wildebeest roamed outside the present boundaries of the park, and it was intended to change the borders to lessen its area.

Nobody can follow these huge regiments of wildebeest and enormous armies of gazelles, and no-one knows where the hundreds of thousands of hooves will march. We were filled with fear and foreboding.

BERNHARD AND MICHAEL GRZIMEK – SERENGETI SHALL NOT DIE

Africa is unimaginably old. Not far from the original site of the game warden's house at Banagi, a few kilometres north of Seronera, stand some of the oldest rocks on earth – two to three billion years old. For much of this time, there were no animals at all. But this ancient land has witnessed other great migrations, and an array of wild animals the like of which will never be seen again. Spectacular as today's herds appear, they are a mere fragment of that golden age of mammals which reached its zenith in Pliocene and Pleistocene times, millions of years ago.

Hanging on the dining room wall at Ndutu Safari Lodge is an enormous skull bearing downswept horns that belonged to a giant bovine named *Pelorovis*. On one side of the ancient beast hangs the skull of a modern day wildebeest, on the other that of a Cape buffalo, both dwarfed by comparison. *Pelorovis* was grazing lush, aquatic pastures around Olduvai Gorge nearly two million years ago, about the time that sabre-toothed cats and giant hyaenas roamed the Serengeti alongside antlered giraffes, gorilla-sized baboons, and elephants with downward-pointing tusks protruding from their lower jaws; a time when perhaps thirty per cent more genera existed.

Further east stand the darkly brooding shapes of the Crater Highlands, a group of extinct and dormant volcanoes in the Ngorongoro Conservation Area. The origins of the Serengeti's animal-speckled plains lie within the history of those far-off mountains that stand watch, dark and menacing, beyond the park's eastern gateway.

Imagine the fiery scenes five million years ago as the land shuddered and heaved, then ripped apart. Lava gushed from the earth's wounds, eventually piling up and coalescing to form the Ngorongoro Crater Highlands. During the explosive volcanic activity that followed, Ngorongoro rose to a height of 5,000 metres, blasting ash and volcanic debris over an area of thousands of square kilometres to north and west. By the time Ngorongoro had spent its fury, other volcanoes were erupting, adding their own ash layers to the plains. The largest of these was Kerimasi, situated at the northern end of the highlands.

Day and night Kerimasi belched huge quantities of fine grey ash into the sky, blotting out the sun and turning the African day into a time of virtual darkness. For more than a million years, layer upon layer of volcanic ash settled over the undulating plains to the west of the highlands, filling in the hollows and smoothing out the contours to form a relatively flat skin of ash across the earth.

OVERLEAF *Africa is unimaginably old . . .*

 Only the granite kopjes stand as a reminder of the ancient land buried beneath . . .

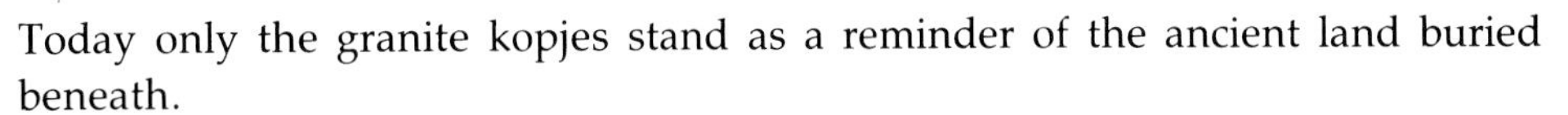

Today only the granite kopjes stand as a reminder of the ancient land buried beneath.

In time rainfall leached the salts from the porous upper levels to form a concrete-hard layer below the surface. This hardpan is impenetrable to all but the shallowest roots, inhibiting the growth of trees, and allowing for the colonisation of the area by an uninterrupted sward of shallow-rooted grasses that reaches out to every horizon for 10,000 square kilometres.

As you drive from south east to north west across the Serengeti plains the volcanic soils gradually become finer grained, annual rainfall increases and the grasses grow progressively longer. Between the long and the short grasses, there is a belt of intermediate grassland dominated by a species of blue-stem grass, *Andropogon greenwayi*, and certain varieties of a drop-seed grass, *Sporobolus*. Close to the sign announcing the rule prohibiting off-road driving within sixteen kilometres of Seronera Lodge, taller stands of grass line the murram track. The wildebeest and zebra rely on these longer grasses to sustain them as they migrate to the woodlands at the beginning of the dry season in late May.

By the time you reach Seronera the Crater Highlands are a hundred kilometres away and barely visible. Here the transition from open grasslands to bush and woodlands is abrupt, reflecting the extent of the alkaline ash soils and their inhibiting effect on trees and shrubs. In Serengeti, a unique blend of soil and climate, plant and animal has combined to create some of the most productive pastures on earth. Nowhere else can you see such a wealth of life.

The volcanic origins of these powdery soils means that they are rich in potassium, sodium and calcium. A pregnant wildebeest could not wish for more nutritious pastures on which to graze and give birth to her calf: perhaps this is why the wildebeest return to these eastern plains each year.

Each morning as the rising sun warms the rugged flanks of the Gol Mountains, hundreds of pairs of Ruppell's griffon vultures preen and squabble on the high rocky cliffs of their breeding colonies. The warming rays breathe life into the thin mountain air, creating buoyant, energy-saving thermals for the birds to fly on. First one, then another, then dozens of vultures spread their giant wings and lift off into the blue to begin their daily pilgrimage over the plains. Soaring and gliding, effortlessly moving from one up-current to another, the vultures search for carrion. Further and further they fly each day, pursuing the herds across the plain and through the woodlands.

Averaging fifty kilometres an hour, these vultures must sometimes travel 160 kilometres in a day to find sufficient food for themselves and their chick. Before night-time settles over the land, many of the vultures return to their nests, crops bulging with meat, clustering in tight feathered groups to listen to the sounds of darkness: a lion's menacing roar, the eerie whooping of hyaenas, the rasping cough of a solitary leopard.

Only the far-ranging vultures can claim to know the secrets of the wildebeest migration. Only they know how far the herds travel, and where to find the vast armies from one day to the next. To solve the riddle of the migration's whereabouts, you must fly.

When Myles Turner arrived at Banagi in November 1956 to take up his position as warden of the Western Serengeti, little was known of the migratory movements of the wildebeest herds, or even how many large animals actually resided in the park. In those early days, Myles led foot patrols into the remotest parts of the Serengeti, pursuing poachers and familiarising himself with every hill and valley of the little-known terrain.

In late 1957, Myles was told that a German professor and his son would be flying to the Serengeti, and that he should prepare an airstrip for them near the park headquarters at Banagi, where Myles lived with his wife, Kay. The professor's name was Grzimek, an extraordinary name that was soon to become a legend in conservation circles.

Bernhard and Michael Grzimek were already familiar with the Serengeti and Ngorongoro Crater, having produced an award-winning film about the area called *No Room for Wild Animals*. Their film protested against the British government's proposal to change the boundaries of the Serengeti, in effect cutting the national park in half so as to accommodate the demands of the Masai. The Grzimeks were horrified by the possible consequences of this decision on one of the few remaining wild places still densely populated with animals.

People had speculated that there were a million large animals roaming the Serengeti, but nobody had ever counted them. As to the migration, no one could say with any certainty how far the wildebeest and zebra travelled each year, or even why they travelled. What factors determined the timing and pattern of their great journey? Until the answers to these questions were better understood, it seemed pointless to finalise the boundaries of the Serengeti. The Grzimeks were invited by the Board of Trustees, at their own expense, to conduct a census of the area and its animals. But time was running out.

The Grzimeks knew from their previous work in the area that it would be impossible to monitor the animals and their movements accurately from the ground. Apart from anything else, the land was imperfectly mapped. Many of the rivers and hills were not shown, and some of the distances were incorrect. During the rainy season, it was often impossible to drive along the few existing roads, let alone venture into the interior across water-logged plains, through swamps and ravines, to skirt round the mountains and rivers that together gave form to this vast wilderness. There was only one way in which the Grzimeks could hope to accomplish their task. They would have to learn to fly.

One hot afternoon in January 1958, Myles Turner heard the sounds of an aircraft approaching Banagi from the east. The bush airstrip was only partially completed, and logs and thornbush lay scattered around. But as Myles was soon

Ruppell's griffon vultures spread their giant wings and lift off into the blue . . .

to discover, the Grzimeks allowed little to stand in their way. The zebra-striped Dornier swooped low over the trees and landed comfortably in the three hundred metres that had been cleared of bush.

The Grzimeks camped near Banagi throughout 1958 and, with Myles Turner to guide them, carried out the first aerial census of the Serengeti's animals. Their original idea had been to try and photograph the whole of the Serengeti from the air and then count the animals in the pictures. Even if practical, it would have cost more than the total profits from their film. And so they were left with only one alternative. They must count each individual animal from the air.

To accomplish their task, the Grzimeks divided the Serengeti into thirty-two districts. When they were unable to find suitable landmarks to act as boundaries, they dropped a paper bag filled with lime and used the white circle as a marker. By flying at heights between fifty and a hundred metres, they could survey a 500 metre strip on either side of their aircraft, enabling them to identify the various types of animals and distinguish young from adults. Usually they flew at two hundred kilometres an hour, but when the ground was thickly populated with animals it became necessary to reduce the speed to just fifty k.p.h. – though not for long, as the engine quickly overheated in the rarefied air. In this way, the Grzimeks covered the entire Serengeti.

But this was not all they did. They collected soil samples for analysis; pressed specimens of grasses for identification to see which species the different animals preferred to eat; immobilised zebra and gazelles to secure ear tags and coloured collars to monitor their movements. They even roped zebra with noose poles and painted them a quince yellow. All so as to know the Serengeti better and to strengthen the case for preserving it.

When it was over, the animals counted totalled 366,980, of which nearly 200,000 were Thomson's or Grant's gazelles and 99,481 were wildebeest. This was only a third of the million large animals that had been thought to reside in the Serengeti.

These figures reinforced the Grzimeks' fears for the future of the wild animals. Poaching, it seemed, was taking an even greater toll on the herds than they had first thought. They now had a more accurate picture of the route followed by the wildebeest and zebra. Their migration took them well beyond the protection of the park boundaries, where the poachers would always be waiting for them.

In 1959, despite much lobbying by eminent conservationists, and the charitable efforts of the Grzimeks, the colonial government changed the boundaries of the Serengeti, excising the eastern plains and the Crater Highlands from the park. These areas would, however, remain protected as the 5,120 square kilometre Ngorongoro Conservation Unit. But the Masai herdsmen and their cattle would be free to roam throughout this area.

But all was not lost – in fact much else was gained. The government abandoned their original intention of removing the entire central plains from the park – an

During the rainy season, it was often impossible to drive along the few existing roads . . .

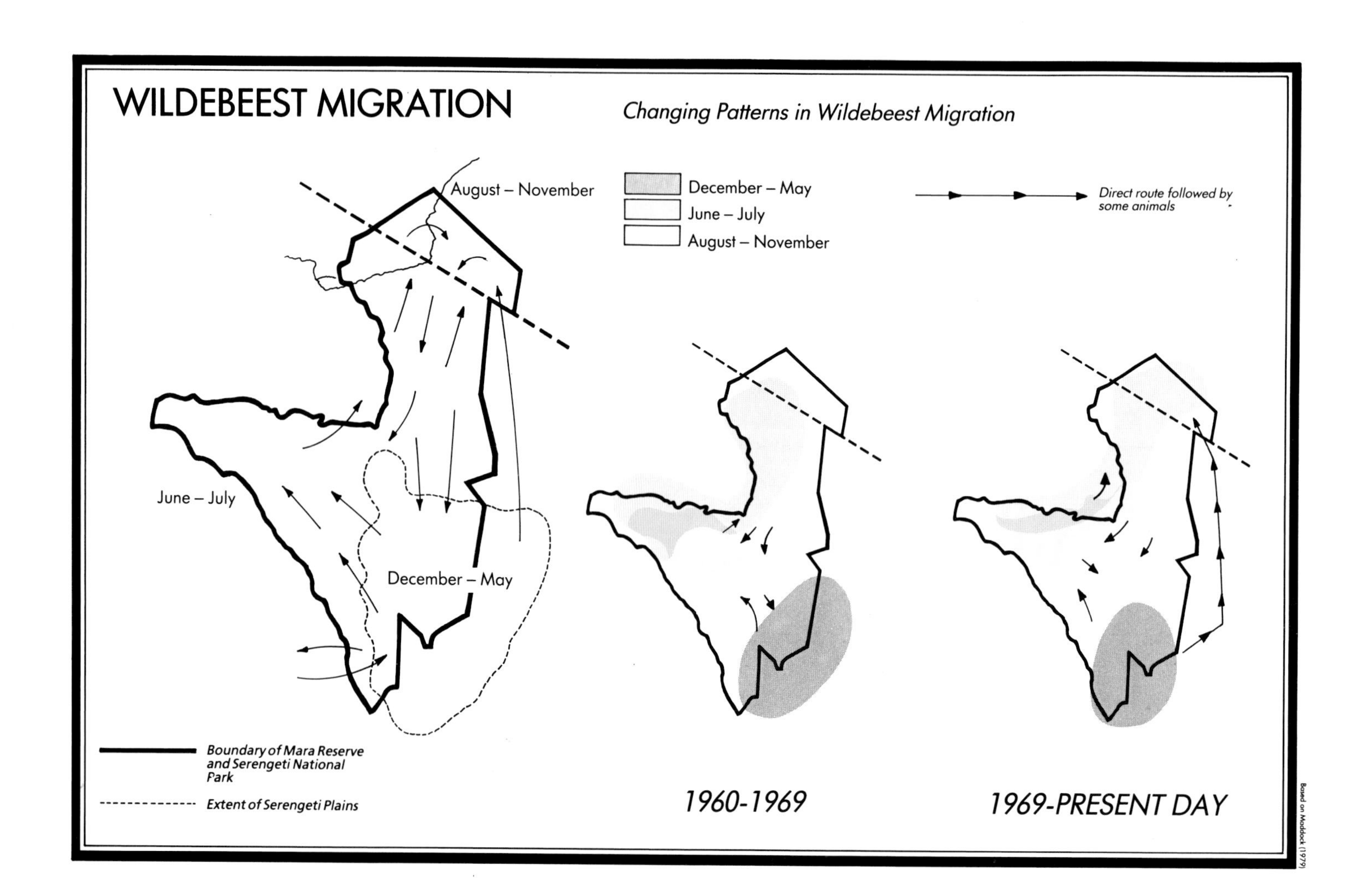

area of 6,600 square kilometres – and returning it to the Masai as grazing land. From now onwards the Masai would be prohibited from grazing their cattle *within* the Serengeti, as they had done for years at places such as the Moru Kopjes. A further 1600 square kilometres were added to the park along the south-western boundary, helping to safeguard the animals on their journey west through the Duma region. And to compensate for the loss of the Crater Highlands, a 3,200 square kilometre block of virtually unexplored woodlands was added to the north. So despite the boundary changes, the *size* of the park remained virtually unchanged.

The addition of the Northern Extension has proved a blessing in disguise. At the time of the boundary changes, the wildebeest population was but a fraction of current levels and the migratory animals had little need for these additional pastures. But in recent years the herds have proliferated. The Northern Extension is now of vital importance. Far from being an inconsequential wilderness, an appeasement for losses elsewhere, it encloses an essential part of the dry season migration route, linking the Serengeti's central plains with the Masai Mara in Kenya.

Since the Grzimeks' pioneering efforts, scientists stationed at the Serengeti Wildlife Research Centre have refined and computerised the painstaking methods of aerial census work. Every two years, during the rainy season, when the wildebeest are massed in a single enormous herd across the open plains, a light aircraft leaves Seronera to begin the daunting task of counting the Serengeti's animals once more. With the help of a radar altimeter set between three hundred and five hundred metres, the pilot is able to maintain a constant height above the herds, enabling colleagues to take a series of vertical photographs with thirty-five millimetre cameras clamped to the outside of the door of the plane or operated through a hatch in the floor. By the time they have flown parallel aerial transects over the entire herd, they have taken a thousand photographs. Using a binocular microscope, assistants then count the animals on the prints, pin-pricking each animal counted so as to avoid mistakes.

When the first photographic census of the wildebeest population was completed in 1961, it revealed a figure of 263,362 wildebeest, nearly three times the 99,481 estimated by the Grzimeks in 1958. Subsequent counts have charted a five-fold increase in the population.

The primary reason for this dramatic increase in the migratory population is thought to be the eradication of rinderpest, a highly contagious viral disease of ruminants. Rinderpest is a kind of bovine measles, known throughout Europe as far back as biblical times. The disease first struck the Serengeti towards the end of last century, having been introduced to Africa by cattle brought in during the colonial period. The results proved catastrophic, and the ensuing cattle plague virtually wiped out Masai livestock as well as devastating the wild herds of buffalo and wildebeest – species which are particularly susceptible to the disease.

Over the next seventy years, rinderpest became endemic in East Africa, periodically ravaging the Serengeti herds. The highest mortality was among calves, particularly during the second half of their first year, when immunity conferred on them by their mother's milk had been lost.

It had always been supposed that cattle contracted rinderpest from their wild cousins. But when the veterinary services mounted an intensive inoculation programme during the 1950s to protect livestock surrounding the Serengeti, rinderpest soon disappeared from the wildebeest and buffalo herds, suggesting that the cattle had been the source of the disease. Once released from the effects of rinderpest and aided by an increase in dry season rainfall during the 1970s, the Serengeti's wildebeest and buffalo populations exploded. When the 1977 aerial count was completed, it revealed the staggering figure of 1.4 million wildebeest, and some people felt that there might soon be two or even three

The wildebeest are massed in a single enormous herd across the open plains . . .

million wildebeest roaming the 30,000 square kilometre Serengeti/Mara ecosystem. But contrary to speculation, the wildebeest population remained at about 1.3 million for the next ten years, held in check by the availability of dry season forage. The latest count in 1986 produced a figure of 1,146,340 – 200,000 less than the previous result. Only when the next count is completed in 1988 will it be known if this is due to the effects of a severe dry season in 1984, an increase in poaching or simply a statistical artefact.

As the wildebeest have long since lost their immunity to rinderpest, any future outbreaks of the disease could spell disaster for the wildebeest hordes, perhaps eventually reducing numbers to 250,000. In such circumstances demands for the land to be turned over to the people would be deafening.

The Grzimeks proved that the great herds could be counted and monitored from the air. Now with the help of space technology, scientists hope to follow the journey of a number of individual wildebeest each day of their life. No longer will biologists need to take to the air or follow their study animal in a four-wheel-drive vehicle simply to monitor its movements. Instead, a transmitter, weighing little more than a kilogramme, will be placed around the animal's neck. This device emits electronic pulses which are picked up by space satellites and relayed back to ground stations. For just three American dollars a day biologists can eavesdrop on their animal's movements via an international computer link-up. A lot cheaper than fuelling an aircraft or buying a four-wheel-drive vehicle.

Many of the oldest secrets of the great migration could soon be revealed: exactly how far an individual animal travels each day; variations in the routes taken from one year to the next; and if there is any interchange between the huge migratory population and the smaller resident herds. It should then be possible to plot the movements of the wildebeest accurately in relation to seasonal changes in climate and vegetation, and to see to what degree the animals are able to predict such changes, enabling them to take best advantage of forage availability.

Biologists have already applied the wonders of satellite telemetry to another of the animal world's great migrations – the Arctic journey of the Porcupine caribou herd – and found it totals twice the distance previously estimated, a journey of up to 3000 kilometres. We may soon know exactly how far the wildebeest travel from one year to the next.

 The African wild dog is an enigma . . .

3 *Life on the Plains*

Wild animals never kill for sport.
Man is the only one to whom the torture or death of his
fellow creatures is amusing in itself.

J. A. FROUDE – OCEANA

For all its ability to stand within five minutes of birth and to out-distance a short-winded lioness, a wildebeest calf will always be vulnerable on the Serengeti plains to the speed and endurance of a coursing predator such as the wild dog. Its only hope of survival against these formidable hunters lies in passing unnoticed, in being only one amongst thousands of others. But for all their legendary killing efficiency, the impact of the Serengeti's wild dog population is barely noticed in relation to such enormous numbers of prey animals.

The African wild dog is an enigma. The mere mention of its name is guaranteed to arouse strong emotions. Admired for their sociable nature and the care they lavish on their pups, wild dogs are still loathed and despised by many people because they kill their prey by disembowelling. Man has always been quick to judge other species, relegating the dogs to a position of social outcast in the animal world.

Today the wild dog is the most endangered predator in the Serengeti. There are probably fewer than a hundred dogs roaming the entire park. Not so long ago such news would have been received with considerable satisfaction. Hunters, farmers, game wardens and even some conservationists took it upon themselves to do everything in their power to exterminate wild dogs. In many parts of Africa, they were shot on sight, or poisoned. In some places they still are. It is only during the last twenty years that naturalists and scientists have been able to present a more balanced view of the wild dog's way of life, illuminating their true role in the natural order of things.

The dogs are nomads, wandering in search of prey over a large home range of up to two thousand square kilometres. For as long as the wildebeest are available, the dogs will try to kill them. But when the migration departs in May and June, the wild dogs living on the Serengeti plains survive by hunting the fleet-footed Thomson's gazelles.

Most animals breed during times of plenty, when food is abundant enough to sustain both adults and young. This is as true of the wild dogs as it is of the wildebeest. The majority of litters are born during the rainy season, between December and May. In years of good rainfall, the wildebeest herds mass on the short grass plains during this period and drop their calves, providing easy prey for the dogs during the three months that their pups remain dependent on a den site. At this time of the year the hyaenas – the wild dogs' major competitors – are often satiated with easy food and their impact is less of a drain on the dogs' resources.

The wildebeest provide easy prey for the wild dogs . . .

The old female had chosen wisely. The Naabi pack's den was situated at the edge of an erosion terrace in the middle of a gently undulating plain. Four kilometres to the north you could see Barafu Kopjes rising from the plains just to the east of Ngare Nanyuki: a tree-fringed stream of alkaline water often used by the passing herds. To the east the blue crest of Lemuta Hill cut across the horizon, and behind it loomed Oldoinyo Gol – the Gol Mountains – where the Ruppell's griffon vultures nest.

The den had been fashioned from a network of old hyaena burrows. Sometimes one of the dogs would rework a partially blocked entrance leading into the den, sending spurts of powdery earth into the air from between its legs. The immediate surrounds were littered with the bones of dead animals – portions of carcasses carried back by hyaenas to feast on at leisure – making the area look like an archaeological excavation. Scattered among the dry bones were the discarded quills of porcupines, other nocturnal visitors who sometimes gained nourishment by gnawing on old bones.

The area adjacent to the den had eroded to form a natural forecourt, creating an ideal enclosure for the female's eleven pups to play and explore in safety. At its northern end the terrace sloped away to a shallow water-hole, much coveted

The old female had chosen wisely . . .

The dogs might wander over and lap briefly at the inviting water . . .

by both predators and prey living out on the plains. The rain-filled pool acted like a magnet for the wandering herds of wildebeest and zebra. The game would gather downwind of the den and stand motionless in a half-circle around the green-fringed water, as if mesmerised, tormenting themselves with the sight of the water, yet knowing from the stench in their nostrils that predators were now in residence. Sometimes a dog would rise from the concealing grass or emerge from a burrow, gazing nonchalantly over the watching herds before flopping down again. At other times, when it was really hot, the dogs might wander over and lap briefly at the inviting water or wade into it and lie on their bellies, sending the skittish herds galloping away in search of a more peaceful drinking place.

The only visitors that seemed relatively undisturbed by the dogs' presence were the birds. With the wonderful advantage of flight they could hold their ground until the last moment before escaping into a dimension that always left the dogs somewhat bemused.

The commonest and most numerous visitors were noisy parties of yellow-throated sand grouse. Early each morning and again in the evening, they would launch from the dry plains like tiny rockets and swing in perfect formation on boomerang-shaped wings towards the pool. Sometimes they were delayed from drinking by the dogs' presence, the earliest arrivals clustering on the plains some metres from the water. But once the boldest had safely drunk their fill, others arrived in their hundreds. As each group departed, the next in line shuffled forward to take their place, as orderly as early morning worshippers taking communion.

Occasionally a chestnut-bellied sand grouse would join the throng, a stream-lined splash of gold, sporting needle-thin central tail feathers. Later when the last sand grouse had vanished, red-capped larks came in small parties to drink and bathe, fluttering their tawny wings and sending a shower of tiny droplets over their brightly coloured heads and backs.

I never did discover the secret of where else the pair of Egyptian geese roamed – or even if they were the same birds. They visited the water-hole for a day, sometimes two, and then disappeared again. When present, they acted like old time residents – bossy and aggressive, strutting around the edge of the pool announcing their territory with strident calls, and only grudgingly waddling aside when the dogs came down to drink or wallow. Perhaps like others of their kind, the geese had laid their eggs in an old hamerkop nest along the tree-lined water-course elsewhere in the Serengeti.

When it rained and the grass around the water-hole stood green and lush, the geese busied themselves feeding, wrenching off the tops of the grass with a deft flick of their bills. Having harvested the ripe grass, they would depart – always with a noisy proclamation – sending eleven inquisitive puppies hurrying back to the security of the den. Large ears cocked, faces turned skywards, the pups watched unblinking as these curious objects passed low over their heads.

. . . noisy parties of yellow-throated sand grouse

When the Naabi pack established their den in mid-January, the wildebeest were already massed on the short grass plains. The rains had been generous throughout December and, as I had hoped, the wildebeest responded by venturing from their dry season holding areas among the woodlands. The plains awaited them, as they had done for thousands of years. In these circumstances the wildebeest appear as one huge group, rolling across the undulating plain with heads bowed to the ground, cropping the grass as they surge forward. Each individual is constantly on the move, though by sheer force of numbers the herds may occupy a particular area for a week or more.

This is the migration people dream of seeing, kilometre upon kilometre of plains country blackened by the wildebeest's presence. And competing for space among them, the other migratory species – zebra, gazelles, and shy parties of eland, wariest of all the larger antelopes.

But the wildebeest could not afford to stay for long in any one part of their range. The rains arrived and departed as scattered thunderstorms. At times, no more than a few drops wet the dust where I waited, while barely more than a kilometre away I could see dark thunderheads stooping to touch the ground, delivering hour after hour of much needed moisture. This pattern of rainfall ensured that at least some part of the plains was always receiving rain. The wildebeest responded by constantly rotating round these natural paddocks: revisiting areas every few weeks to recrop the grasses when they were at their most nutritious.

OVERLEAF *This is the migration people dream of seeing . . .*

While living in the Mara, I had grown accustomed to a landscape dominated by the russet colours of waist-high red oat grass. But even during the height of the wet season, when the Mara lies buried beneath an ocean of long grass, the Serengeti's short grass plains do not sprout more than about a few centimetres from the earth. I had never seen grasses quite like these before.

If you abandon your vehicle for a moment and press your face close to the ground you get more than a wildebeest's perspective of the plains: you find a mosaic of little grass cushions, as neatly arranged as artificial turf. The grey, powdery soils peer from between these earth islands, shielded from the erosive effects of wind, rain and grazing by a dense mat of herbaceous plants. The grasses are shallow-rooted perennials able to grow unimpeded by the tree-defying hardpan and ideally suited to the alkaline soils. Their roots are covered with minute hairs which absorb every drop of condensation accumulating between the soil particles during the chill nights. Even in the driest spells, the life of the sward is assured – stored away in the roots, awaiting the next rain. In this way the grasses can survive the withering effects of drought, yet erupt within hours of rain to nourish the wet season deluge of more than a million wildebeest.

Physically the short grasses look quite different from those found in the long grasslands. They are dwarfed by comparison, bearing small fine leaves and prostrate stems: a protective response to the thousands of hungry mouths that prey on them throughout the wet season. It is grazing that keeps the grass short, and grazing that helps keep the pastures growing. Proof of this can be found inside the scientists' odd-looking exclosures set among the grasslands. These square, fenced-off areas prevent the animals from feeding on the enclosed grasses. Here you can see that left ungrazed during the rains the grasses would soon become mature and flower, growing to heights of up to sixty centimetres. Without the grazing herds, the composition of grass species found on these plains would change, and the taller grasses would crowd out the more prostrate forms in the battle for light. Thus the wildebeest actually help to create the grazing conditions they prefer – short green grass yielding a high leaf-to-stem ratio. By constantly mowing the grass in a massed feeding front, they prevent less palatable species from flourishing.

Surprisingly, perhaps, the presence of so many wild animals does not damage the plains environment. The wildebeest never stay long enough to do that. As they sweep from one part of the grasslands to another they constantly enrich the thin soils, helping to recycle the nutrients before moving on. Everywhere one looks there are dark piles of dung and urine: vital natural fertilisers. Even as the animals clip the grass, growth hormones pass from roots to shoots, promoting rapid re-growth. And the saliva of the herbivores acts as a stimulant to the grasses. Plants and animals survive in a state of dynamic harmony, allowing both to prosper.

The wildebeest slipped away to the nearest green area . . .

By early February the great mass of the migration were headed for the Gol Kopjes, where the rain had fallen a few days earlier. As they disappeared over the horizon I remained with the wild dogs, leaving the wildebeest to slip away to the nearest green area, close to the wooded country around Lake Lagarja in the south.

Soon the eastern plains looked empty. A chill breeze blew each morning, sucking back the moisture given to the plains. Friends told me that the wildebeest had moved many kilometres away, marching across the land, sometimes walking, often running. They must travel quickly if they are always to be within reach of food and water.

Within days – hours even – the zebra moved in, and from Barafu to Gol the air was filled with their braying, dog-like calls. Soon they were scattered far and wide across the plains. Wherever I looked I now saw zebra. Not feeding together in a single mass like wildebeest, but moving across the plains in loose aggregations numbering hundreds of animals. Within these herds I could pick out distinct family groups, consisting of a few females and their young, jealously guarded by a single stallion. Morning and evening, the zebra crowded around the water-hole, prevented from drinking by the indolent presence of the wild dogs. But it soon became apparent that the zebra had little to fear from the Naabi pack, though it was not through lack of effort on the dogs' part. Each evening as the temperature dropped, they would stir themselves, performing the elaborate greeting ritual that always precedes a hunt. Then they were off, trotting loose-limbed across the plains in search of prey.

The three yearling dogs . . .

The three yearling dogs, born to the old female the previous year, were still relatively inexperienced killers. They often tried to initiate hunts of fully grown male wildebeest or zebra – animals more than ten times the weight of a wild dog. Usually the adult dogs simply ignored these exuberant diversions, unless it looked as if success might be imminent.

Zebra families are highly co-ordinated in their defence against wild dogs, bunching together and preventing the dogs from reaching vulnerable foals. The stallions fall back, biting and lashing out with their hooves at any dog that dares to try and close in on one of their number. None the less there are packs of wild dogs that have proved to be highly successful hunters of zebra. Such packs contain males experienced in the art of immobilising a struggling zebra by leaping up and grabbing it by the nose and upper lip. Once it is secured in this fashion, other pack members can safely move in and disembowel their prey. But it is a difficult and dangerous pursuit and most wild dogs rely on smaller and easier prey.

Even when the Naabi pack managed to isolate an injured zebra, they still found it impossible to get past the animal's flailing hooves. The yearling dogs would repeatedly lunge forward to try and grab a leg or tail, often narrowly avoiding serious injury. But eventually they were forced to abandon the zebra and turn their attention to the Thomson's gazelles. The tommies were faster but far less intimidating prey, which would help to feed the Naabi pack and their pups until the wildebeest returned.

In 1986 I hardly saw another vehicle out on the plains. The rebirth of tourism was still in its infancy in Tanzania, and when the Naabi pack denned only a few kilometres to the east of the main road between Naabi Hill and the Simba Kopjes, they went virtually unobserved except for visits by scientists from the Serengeti Wildlife Research Centre. But when they denned at Barafu in January 1987, word of their whereabouts soon got around and visitors travelled great distances to see them.

Part of the reason for all the attention was the enormous increase in tourism to Tanzania. Once again the lodges were full to bursting and there was a new and welcome spirit of enterprise at the Seronera and Lobo Lodges. The rains were heavy and well-distributed in the north and west from December through to March, and popular game-viewing areas such as the Seronera Valley and Moru Kopjes – which support a good variety of resident game year round – were soon enveloped in a shroud of waist-high grass. Game-viewing became more difficult in these areas: roads were at times impassable and animals hard to find, let alone photograph, in all that grass. It was time for the drivers to visit the plains further south.

With the wildebeest calving out on the plains, and cheetah accompanied by their cubs hunting daily in the vicinity of the Gol Kopjes, visitors to the Serengeti were seeing it at its unrivalled best. Having once ventured as far as Gol, tour

. . . in the vicinity of the Gol Kopjes . . .

drivers were only too happy to extend morning game drives, and continue north east to Barafu so as to show visitors something as rare – yet visible and active – as a family of wild dogs at a den site. It was cheering to see so many people enjoying the sight of the old female nursing her pups; to witness the sharing of food between adults and young through the extraordinary process of regurgitation. The wild dogs need all the friends they can muster.

Barafu had never received so much attention. Arguments soon raged as to exactly where tour operators should be permitted to set up mobile camps, and whether it was better to follow existing tracks or to establish a new one on each passage over an area as fragile as the short grass plains. The consensus of scientific opinion agrees with the wooden sign erected next to one of the Gol Kopjes: following car tracks destroys the vegetation.

The most visible tracks were those left by heavy, double-wheeled trucks. One journey through an area by such vehicles, particularly while the ground was wet and soft, left deep depressions that might later erode into an ugly scar. The park authorities have therefore decided to limit the movement of such vehicles – used to ferry equipment into camp sites, and to collect water and firewood – to the main roads. In future camps will have to be equipped by smaller vehicles, and

The wildebeest and their calves were moving through the area . . .

no one will be permitted to camp at Gol and Barafu.

I stayed with the Naabi pack and their latest litter of pups throughout February and March. Sometimes, after the adults had returned from a hunt, regurgitated meat for the puppies, and settled down to rest through the hottest hours of the day, I retreated to the shade of the nearest kopje. From there I could still keep watch with binoculars in case something dramatic that I wanted to photograph seemed likely to happen. And it often did when the wildebeest and their calves were moving through the area. The welcome sight of those small silhouettes emerging over the horizon with their mothers was guaranteed to rouse the dogs to action. Especially now that there were eleven extra mouths to feed.

Barafu Kopjes sprout from the dry skin of the plain in a tight cluster, joined together by a central spine of rock set deep beneath the surface. The surrounding plains are criss-crossed by deep ruts charting the path taken by generations of wildebeest as they plod back and forth across the land. During rainy periods, water collects in a network of shallow depressions on the bony surface of the kopjes. This provided me with a valuable supply of water that I could scoop up into a bucket so that I might wash myself and my clothes.

What luxury! In 1986 water was scarce on the plains, confined to a few muddy pools, used as both wallows and latrines by the dogs and hyaenas. The wildebeest spent much of the early part of that year in the woodlands of the Maswa Game Reserve, to the south west of the Serengeti, loitering as near to the plains as green grass would allow. That year fuel was even more precious than water, and I was often far from the nearest kopje which might hold the rainwater long enough for me to use it.

On reaching the largest of Barafu's outcrops, I circled my car around whatever shade I could find, trying to keep one step ahead of the sun. In the same way I rotated my precious store of Kodachrome film, moving it throughout the day until I had exhausted all of the secret cool spots that I had discovered in my car.

I relished the opportunity to escape from the confinement of my vehicle, to stretch my legs and revel in the vastness of the land. For exercise and a change of pace, I daily climbed the steepest rock face of the largest kopje, convincing myself that it was the north face of the Eiger – though my mountain was barely ten metres at its highest point. I always moved cautiously on my walks around the ancient stone islands. I knew that at times the kopjes acted as home or resting place to lions, leopards and spitting cobras. The possibility of their presence created a pleasing sense of tension, recalling how it must have been a million years ago when early man competed for food alongside the great predators – physically frail in the face of such awesome power – seeking to survive through the gift of his brain.

There were signs of life everywhere. Rocks splashed white, where vultures and eagles had perched amid the scats of other unseen wanderers: baboons, lions, and the bone-white droppings of hyaenas. I had heard of a shy leopard that sometimes ventured across the treeless plain between the head waters of the Ngare Nanyuki water-course and Barafu to its east. Though I never saw him myself, I found clues to his activities.

Crowned plovers screamed out a warning whenever I appeared over the rocky horizon. And brightly coloured male agama lizards violently bobbed their heads in my direction, proclaiming their tiny territories before scuttling away to the safety of a concealing crack. Gazelles looked up, issuing nasal alarm calls of their own at the sight of the most dangerous predator of all.

From the top of the kopje, I had a perfect view across the plains. Below me a pair of huge lappet-faced vultures, largest and most powerful of all the six species found in the Serengeti, tended their crudely constructed nest: an untidy pile of whitewashed sticks dumped carelessly in the top of a lone acacia tree. These birds behaved quite differently to the Ruppell's vultures nesting in the Gol Mountains to the east. A pair of lappet-faced vultures fiercely defend a fixed territory year round, rather than pursuing the path taken by the migration. Having once located a carcass, the lappet-faced vultures use their huge beaks and hefty feet to ward off the griffons, who tend to dominate a carcass through sheer force of numbers. After a few visits, the vultures seemed to accept my presence, and I could watch them through binoculars as they fed their chick.

The roof of the kopje is adorned with all manner of bristly succulents, aloes and wild cucumbers, their colours merging perfectly into muted tones of yellow and green. In places, huge slabs of granite have broken free from the parent rock, as if a craftsman had split and fractured them into enormous stone

There were extra mouths to feed . . .

The possibility of the predators' presence created a pleasing sense of tension . . .

sculptures. Some lean and perch at impossible angles, so finely balanced, it seems, that I hardly dared touch them for fear of sending them crashing. Walking over the rocks created a hollow, resonant tone, like the sound of a distant drum, reminding me of the 'gong rocks' at Moru Kopjes where early man had played music.

On one visit I sat unnoticed as an elegant cheetah, heavily pregnant, stepped delicately across the plain below me. The gazelles parted to let her through, then turned to follow in her wake, snorting in protest. Suddenly she broke into a trot. Within seconds she was galloping, singling out a mother and fawn. The chase was long and fast, but as she closed to within a couple of metres of the young gazelle, it turned aside sharply, leaving the cheetah panting in the midday heat. This same cheetah also hunted wildebeest calves, though for most of the year it would be the gazelles that provided her with food for her cubs. When the gazelles moved to the woodlands at the onset of the dry season, she would abandon this area and follow. They were her favourite prey.

Every day I climbed the kopje to watch as the scene transformed – some days dry as a bone, others wet and green. Calm or windswept, hot one day, cold the next, the permutations were endless. Whenever the wildebeest moved away from Barafu I noticed the dainty brown and white gazelles as if for the first time. The grazing pattern of the wildebeest reduced the grasslands to a neatly clipped lawn, exposing the nutritious shoots and scattered herbs which the gazelles prefer.

Each day, the wind blew, drying the plains and reclaiming the precious surface water from the rock pools. By the middle of March, the ground looked scorched, though the grass roots still survived unimpaired beneath the rock-hard surface, waiting for rain. At times the heat was almost unbearable, and for the first day in a week huge clouds barrelled across the blue sky, casting thick black shadows over the plains. I longed for rain to refill the pools so that I could wash again and feel cool water on my face. And so that the wildebeest would once more return to Barafu in their thousands.

I awoke long before sunrise with the sound of a distant vehicle buzzing in my ears. This was the last thing I expected to hear: Barafu was many kilometres from the nearest camp, and besides, people are prohibited from driving in the park during the hours of darkness. The sound soon faded to a distant hum as I dozed off again. Perhaps it was scientists from the Research Centre conducting a night-follow of predators. But later, as the sleep cleared from my head, I realised that it was not a vehicle at all. It must be the wildebeest returning to the plains in response to the rain that had fallen during the last week and a half. I quickly started the engine and headed south, leaving the Naabi pack huddled around the entrance to their den. They did not even stir as I drove away.

Viewed from an aeroplane, or the top of the tallest kopje, the plains look like a single relatively flat expanse of land. But this is an illusion. The ground dips and swells like the waves rolling across the surface of a great ocean. Driving over a rise I was suddenly among the wildebeest herds: the most immense visual spectacle one could ever hope to see. No words can convey the impression of so many animals seen together. It is impossible to calculate what percentage of the Serengeti's 1.3 million wildebeest you can take in with a single glance. But it is a sight you will never forget.

And so it continued throughout April and early May, the wildebeest, zebra and gazelles, all constantly moving in the wake of the rains. There were times when it poured, when the plains were so wet that it was virtually impossible to avoid grinding to a halt in the mud. The rutted tracks leading from Barafu to Gol Kopjes ran riot with a flood of muddy water. But within a few days the sun emerged again in a brilliant blue sky and everywhere looked green and inviting.

The wild dog puppies were now old enough to follow the adults: eleven boisterous juveniles gambolling in the wake of their doting relatives. Freed at last from dependency on a den site, the Naabi pack could resume its nomadic ways. It was time for me to bid the dogs farewell until next year, when I hoped to take up their story again. Trying to follow them on a daily basis was now impossible: by tomorrow they could be many kilometres away. I wondered how they would survive the rigours of the dry season. There were lean times ahead of them. Soon every last wildebeest would have fled these plains. The muddy water-holes were already retreating. The herds seemed restless, as if they could sense that the long march to the woodlands was soon to begin.

There were times when it poured . . .

 The wildebeest turned their backs on the stinging wind . . .

4 *The Long March*

Nowadays we don't think much of a man's love for an animal; . . .
But if we stop loving animals, aren't we bound to stop loving humans too?

ALEXANDER SOLZHENITSYN – CANCER WARD

At first it seemed innocuous enough, nothing more than a pleasing breeze each morning, bringing cool air to temper the heat. But soon the dry north-easterly winds continued unabated: gusting night and day, whipping the powdery soils into towering dust devils. The wildebeest turned their backs on the stinging wind and, like a river in flood, began to stream off the plains, drawing to them tributaries of animals from every corner of the land. Dividing, then merging again, they passed through the kopjes of Barafu and Gol; damming up in black clusters to slake their thirst wherever they could still find water seeping from the base of the granite islands. Others had already headed north towards Loliondo, within a few days' walk of the Mara, bequeathing the open plains to the gazelles.

Groups of lions and solitary leopards looked down on the passing herds from their hiding places among rock and tree: for as long as the animals continued to pour over the horizon the predators would have no difficulty finding food. Hyaenas dogged the rutted pathways cut by the wildebeest. Many of these shambling hunters would soon abandon the barren eastern plains and move north west to establish dens at the edge of the woodlands, commuting for days at a time between the Western Corridor and their den sites to prey on the migratory herds during the dry season.

The prodigious appetites and protein-rich diets of the predators enable them to endure both feast and famine. If need be, lions and hyaenas can survive for days without food. Not so the wildebeest. They must forever wander in search of green grass and water. By tomorrow they could be far away, forced to graze for sixteen hours a day to obtain sufficient food – the same amount of time that a predator spends resting.

Lions and leopards are unable to exploit to the full the abundance of food represented by the migratory herds. With highly dependent young, the cats cannot afford to travel too far afield in their search for prey. If they did, their cubs might starve or be killed by other predators. They cannot go off and hunt and then return with a belly full of meat to regurgitate for their young, like the wild dogs. Instead, lions and leopards live as residents within a permanent home range. Though some lions venture out on to the plains during the wet season, most do not follow the herds on their journey north. Those who do are nomadic lions who no longer belong to a pride.

During the dry season, the plains are an inhospitable place for the larger carnivores – no water, no shade, and very few prey animals – just wind and dust. For this reason most lion prides occupy territories in the woodlands, towards which the migration was now headed. But in recent years the Serengeti lion population has increased – particularly out on the plains. It is thought that a rise in rainfall during the dry season, leading to an increase in the numbers of

Groups of lions looked down from their hiding places . . .

resident prey species such as warthog and topi, has enabled this to happen. Many of the lions one now sees resting in the shade of a kopje or crouched on the open plains are not nomads seeking to survive by occupying the harshest and most marginal of lion environments. They are members of prides or nomads attempting to establish prides. But many do not succeed.

Hyaenas are still more flexible. Though they occupy a particular range, which they will defend against other hyaena clans, these areas shift markedly in time and space, reflecting seasonal changes in their food supply. During the dry season, hyaena females are forced to leave their young cubs to go and search for food. Their journey invariably takes them into the Western Corridor where large migratory herds of wildebeest and zebra linger during the early part of the dry season. Having made a kill or while scavenging, hyaenas bolt their food, rapidly

consuming as much meat as possible. This is the most effective way of competing with other hyaenas, and the lions who may already be running across the plains to steal the kill.

By the time a female arrives back at her den she may have travelled nearly eighty kilometres. By now her udders are bursting with milk. Though they eat meat whenever they can get it, hyaena cubs must suckle throughout their first year – sometimes until they are nearly eighteen months old, by which time they are virtually fully grown. It is only then that they are large enough, and strong enough, to kill for themselves, and to compete among the mill of bodies that quickly accumulates around a fresh carcass.

But even the adaptable hyaenas are unable to raise cubs and follow the herds later in the dry season. Once the animals move deep into the Northern Extension and on into the Mara, the hyaenas must rely on resident prey. Only the vultures can keep pace with the wildebeest and zebra, can travel 160 kilometres in a day and still return with sufficient food for their young.

Early one morning, not far from Simba Kopjes, I met the great armies of wildebeest marching through chest-high stands of ripe grasslands that had remained virtually untouched during the rainy season. I pulled off the road and watched as wave after wave of animals crested the rise. Within minutes they were lost to the naked eye.

Hyaena females are forced to leave their young cubs to go and search for food . . .

The migratory herds plunge onwards . . .

At times the migratory herds plunge onwards for kilometre after kilometre, thundering across the plains, streaming over hillsides and fording any torrent that stands in their way, barely pausing for breath. Through binoculars I could see tens of thousands of other wildebeest passing through the majestic Moru Kopjes at the edge of the woodlands in the west, where the Masai used to graze and water their cattle. Beyond them stood the Itonjo, Nyaraboro and Nyamuma Hills, which with others comprise the southern and central ranges. The hills stand like sentries across the wide mouth of the Western Corridor, divided in places by broad valleys and deep ravines. Snaking between these low hills are the Duma, Mbalageti and Grumeti Rivers which meander through the funnel-shaped corridor.

Together the hills and water-courses give form to the westward movements of the herds, channelling them from the plains through a series of rocky gateways leading towards Lake Victoria. The surrounding country is relatively flat acacia woodland, with tall stands of whistling thorn giving way in places to open plains, such as those of Ndoha and Dutwa, Musabi and Ruana, where the wildebeest spread out to feed.

The general direction taken by the wildebeest when they first leave the plains mirrors the rainfall gradient, which increases from south east to north west and is strongly influenced by the towering presence of the Crater Highlands. The short grass areas in the south east lie directly in the rain shadow of the highlands and only receive 500 millimetres of rain annually, almost all of which falls between December and May. Rainfall in areas to north and west is more evenly distributed and may be as much as 1200 millimetres annually, some of which is generated during the dry season by evaporation from Lake Victoria and by thunderstorms caused by grass fires. This dry season rainfall creates a green flush on grazed or burnt areas which is sought after by the migratory herds.

Some years the transition between wet and dry seasons is particularly marked. In these circumstances the stage is set for a truly dramatic exodus of animals, with hundreds of thousands of wildebeest departing for the woodlands at the same time. In fact the term 'the migration' is most commonly associated with this phase of the wildebeest's journey, when the zebra and Thomson's gazelles are also on the move. But whatever definition one applies to the word migration – a change of abode, a seasonal movement – it is misleading to refer only to this one part of the journey as 'the migration': the wildebeest's nomadism is simply a question of degree. I have used 'the migration' as a general term to describe the migratory population of wildebeest – not to pinpoint a particular phase of their year-round journey. The trek from plains to woodlands is only more spectacular because this is the time when the herds must travel furthest and most rapidly to where water and grazing are to be found. The animals are distancing themselves from the famine of the dry season plains and moving quickly to areas in the north and west where food and water are readily available.

. . . streaming over hillsides . . .

They even rut while migrating . . .

Wildebeest are dependent on water. They must either drink regularly or find grazing with a high moisture content. So it is not surprising to see the herds hurrying to the nearest areas where they can find water as the plains dry up. Three of the most favoured watering places on the march west are Oldoinyo Olobaiye, which forms the head waters of the Simiyu River; the Moru Kopjes where the Mbalageti River rises; and the Seronera itself.

Nature has designed the wildebeest to be able to move rapidly and economically over vast distances in search of food – despite their strange looks, they *are* the right size and shape. It takes no more energy for a wildebeest to run a particular distance than it does to walk it. No wonder wildebeest seem in such a hurry – this way they can take advantage of the scattered distribution of green grass, yet still remain within range of water. During the dry season it is a race against time. They run to survive.

The griffon vultures are also adapted to forage over great distances, but they soar on thermals to buy time and conserve energy – wildebeest canter. While the wildebeest chase the green grass, the vultures fly in search of those who fail. Both their lives are held in balance by their ability to travel over huge distances in search of an unpredictable food supply. Their success can be measured by the huge numbers of each species that live on the Serengeti.

Nearly every facet of the wildebeest's behaviour seems to have been designed to save time – compressed and abbreviated through necessity. But how else could they accomplish so much so quickly? They even rut while migrating. And calves gain their feet within minutes of birth so that they may survive in the open and keep up with the ever-moving herd.

One of the great mysteries of the wildebeest migration is the manner in which the animals synchronise the peak in calving, enabling them to bring forth young at the time most suited to their survival. This is only possible if females ovulate at the same time each year, and males restrict their sexual activity to a distinct rut in May and June. The time of mating is critical.

The way in which this complex chain of events is co-ordinated is unclear. Seasonal changes in a variety of environmental factors are known to trigger the production of sex hormones in many animals. Bird species in North America and Europe often begin breeding in response to the lengthening hours of daylight in spring. In equatorial regions, where day length varies little, it is the onset of the rains which acts as the cue. Social stimulation, or other behavioural mechanisms, may perform the same function.

The exact timing of the rut is thought to be triggered in some way by the lunar cycle, working in conjunction with seasonal changes in other environmental influences. But whether the moon's influence stimulates ovulation in females, or induces rutting behaviour in males is unknown. Perhaps it helps to cue both, or perhaps oestrus in females causes the males to rut.

The rut is a refreshing contrast for the human observer. Wildebeest society generally does not favour expressions of individuality: survival is enhanced by immersion in the herd. Stand out and a predator is more likely to select you for its next meal. It seems to be a drab life, one spent among countless thousands of other similar beings; a life concerned primarily with obtaining sufficient food while avoiding being eaten. When not advancing in an irrepressible feeding front, wildebeest are surging over the horizon in search of fresh pastures or massing at a water-hole to drink. There seems little time for anything else.

In years of good rainfall, both males and females have feasted on a steady supply of green forage and are in prime condition to withstand the stress of reproduction with large reserves of energy stored in the form of fatty deposits. The bulls look big and impressive, their coats shining grey with dark fingers of coarse hair streaking their muscular necks and shoulders.

Where before it appeared to be just so many animals – as if every wildebeest were created a clone, functioning as an identical image of the animal standing next to it – a new pattern of organisation now becomes apparent. Fuelled by rising levels of the male hormone testosterone, contenders for territories begin to emerge from the staid bachelor societies to which adult male wildebeest belong for much of the year. The bulls cast off their normally placid mien and start to challenge each other: chasing wildly through the herds, bucking and kicking up their spindly legs, tossing their heads and generally doing everything possible to draw attention to themselves. For the moment the females and younger animals continue peacefully feeding, ignoring the bizarre activities of the bulls.

By the time the rut begins in earnest the animals are already moving en masse from the plains to the woodlands. They are now at their most concentrated: males and females are in the same place at the same time, and have no need to seek each other out. This helps to synchronise the bulk of sexual activity to within the space of a few weeks, ensuring that the next year's calves are born at the optimal time. By then the herds will have returned to the plains and, rain permitting, have access to high quality forage for a month or so before the birth of the calves.

The rutting of the wildebeest is the closest approximation to organised chaos that I have ever witnessed. So much is happening that you hardly know where to fix your gaze. The noise is deafening. Yet the frantic activity of the bulls is far from indiscriminate. To mate they must acquire territory. Each year the older bulls – usually at least three or four years old – do battle for temporary possession of a tiny patch of turf. Fighting is ritualised as a harmless jousting display, designed to select the males most fitted to pass on their genes without unnecessarily injuring their rivals. Territorial bulls constantly perform this ritual with neighbouring bulls, galloping out to meet each other, then dropping to their knees, head to head. Most of the time they do not even make contact. But on occasion they push and spar, horn against horn, the dust rising in clouds around them. Then they are up and off, heads held high, moving away with their characteristic rocking-horse gait. So frenzied do the bulls become that they even gallop aggressively towards a passing car, determined to confront any moving object that impinges on their territory.

There is nothing visible to mark the boundary of each bull's territory. Wildebeest bulls *are* the territory. The only symbols of ownership are the bulls themselves and their stamping ground: a patch of grass-torn and earth-scattered turf where the bulls urinate and defecate, scrape and hook, smearing scent from their inter-digital and pre-orbital glands on to the bare earth.

It is only when seen from the air that the structure of the migrating herd becomes more obvious. At the outer edge a moving wave of wildebeest can be seen with heads bowed to the ground, feeding. Many of these are bachelors – adults, and groups of yearlings that have recently left their mothers – organised into separate herds within the overall mass. Behind this wall of animals a mosaic of territories is revealed – scores of them – each little more than the area defined

The Long March

. . . the closest approximation to organised chaos that I have ever seen . . .

The animals lie down on the open plains . . .

by a tight knot of twenty to thirty females and their calves, guarded by a single territorial bull. The bachelors are constantly forced to the edge of the rutting grounds, if they are to avoid harassment by the territorial bulls. When the bachelors stray too close, the bulls quickly move out to prevent them mingling with the females.

Whenever possible the bulls try to increase the number of animals under their control. A male suddenly sees the chance to pirate one or more females from the edge of an adjacent territory – which may only be metres away – galloping forth to round up any females that stray. In so doing he risks losing females from his own territory and will certainly be challenged by other territory holders. The slightest lapse in attention permits some of the females to move on, which they periodically try to do. When this happens the bull frantically attempts to prevent a mass exodus of all the females from his territory. Being naturally gregarious, when some wildebeest leave, others try to follow. Consequently the males are unable to keep the same females on their territory from one hour to the next – sometimes even from one minute to the next. In this way females are impregnated by a number of territory holders during the course of the rut, and territorial males have the opportunity of covering large numbers of females.

At times, when it is very hot, the herd pauses to rest, and then the animals lie down in the open plains, or cluster in tight groups under the shading umbrella of acacia bushes, lying flank to flank or rump to rump. In this highly sociable manner, wildebeest distinguish themselves from the majority of other antelope who, like many animals, cherish a need for individual space. Meanwhile the territorial bulls remain vigilant, though their activity is somewhat subdued for as long as the females remain quiescent. An air of calm settles over the herd. But wildebeest are rarely static for long. I watched the females depart, forcing the territorial bulls to abandon their territories or be left behind, which they cannot afford if the grass has gone. So the males gallop along, chasing wildly through the migrating herd until the pace slows enough for them to re-establish temporary territories. Some race ahead and mark out territories in the path of the oncoming masses. In areas such as the Ngorongoro Crater, supporting a resident population of wildebeest, the bulls stake out permanent territories, sometimes maintaining them year round, plainly an impossibility for the migratory animals whose social life is geared to the demands of living on the move.

This unrelenting activity on the part of the bulls leaves little time for them to feed. They are far too busy displaying and defending their territory, mating with as many receptive females as possible, burning up the energy reserves stored during the previous months spent foraging on the plains. The bachelors meanwhile are able to maintain their condition and provide a reservoir of fit animals to replace exhausted territory holders. Once ousted from his territory a bull can rejoin a bachelor herd and regain condition before the energy-sapping dry season reaches its peak.

Fighting is ritualised as a harmless jousting display . . .

Later that same day, I drove back towards Seronera, pausing every so often as the herds swept back and forth across the murram road. Soon I was completely surrounded, marooned among thousands of wildebeest as they spread throughout the whistling thorn country to the west of Seronera Lodge. To the north the great wall of rock called Nyaraswiga rose three hundred metres above the plain, standing blue and solemn against the evening sky. As darkness closed around me I stepped carefully from my vehicle. I wanted to savour the immediacy of all those animals and feel the warm earth beneath my bare feet. The still night air was filled with the incessant grunting and honking of the wildebeest bulls, their territorial routine unbroken by the transition from day to night. They were spacing themselves by sound and the eerie light cast by the ripening moon.

Suddenly, the noises changed, building into a deafening crescendo of nasal alarm snorts and strident honkings. Danger stalked the darkness. Soon innumerable hooves beat a rhythmic tattoo on the dry ground as the animals panicked, sweeping wildly around, first one way and then another – frantically trying to escape the predators which had crept among them; standing for a moment to assess the danger, then once more blundering away. Before long the wildebeest were coughing and sneezing on the thick dust churned up by thousands of their own feet. And somewhere nearby the predators were already feasting. Later, as I lay in bed under a star-studded sky, I listened to the hyaenas whooping and giggling. And far off, the fading grunt of a lion receded into the night.

. . . the fiery red of flame lilies . . .

I awoke next morning to find that the majority of the animals had moved on. Suddenly, I seemed to be losing the migration. It had been easy to follow the wildebeest as long as they remained visible on the short grass plains. Now, as May drew to a close, they were daily slipping away into that vast expanse of bush country Myles Turner had once described to me. Part of the problem was undoubtedly the limited visual horizon afforded by my vehicle. No wonder Alan and Joan Root, the wildlife film makers, had opted for the tranquil charms of a hot air balloon while making their epic, *The Year of the Wildebeest*. I had already tried coaxing my shattered vehicle up the side of the nearest rocky hill to obtain a better perspective on the great herds. And one morning I was rewarded with a breath-taking view of the columns of wildebeest tracing a maze of patterns across the plains below, backlit by the early morning sunrise. But I felt increasingly imprisoned behind the wheel of my green tin box, engulfed by the tractor-like drone of the engine and the stench of diesel. My spirit longed to join the herds on their journey, to transcend time and space.

Accompanied by a ranger, I left the last stragglers behind me and drove to the head of a valley, surmounted to north and south by steep ridges. I parked the vehicle in the shade of a cool acacia bush and climbed skywards. Gradually the scene widened into a spectacular eagle's eye view of the Western Corridor. I was engulfed by the feeling of space, of endless country untouched by man.

The euphoria did not last for long. Soon my binoculars picked out settlements clustered hungrily along the park boundary. To the north a grass fire spewed acrid smoke into the dry air, fanned for kilometres around by the strong easterly winds. Another even larger fire burned red hot further to the west, ignited perhaps by the casual toss of a cigarette butt from a passing vehicle. Or was it the work of poachers deliberately scorching away the drying remnants of the previous season's growth, drawing the unwary game animals on to the tempting green flush that would erupt within days of showers, and luring them to an agonisingly slow death in the choking grip of wire snares or as victims of poison-tipped arrows? Sometimes there were guns and vehicles. Perhaps it was, after all, only an illusion of paradise.

I followed a narrow trail where buffalo and elephant had trampled the long grass ahead of me and flowers sprouted colourfully among the ankle-tugging grass tussocks – wild hibiscus splashed white and purple, the fiery red of flame lilies and tiny kidney-shaped vetches. I carefully nudged a football-sized dropping left the previous night by an elephant, revealing a horde of dung beetles that had quickly gathered to process the nutritious find. The beetles frantically moulded the dung into transportable balls of food in which females would lay their eggs. Later perhaps, a black-backed jackal would come sniffing along this game path. It might pause for a moment before hurriedly clawing into the dry soil, breaking open the hollow dung ball and eating the young beetle – the waste products of one animal ultimately providing food for many others.

Flat-topped boulders worn smooth with age beckoned as I surmounted the

. . . ranging from giraffe . . .

rise, offering a vantage point from which to enjoy the view. Far below I could see the wildebeest, arranged like strings of black pearls across the bare bones of the country. Scattered among them were family parties of zebra adding a splash of colour to an otherwise sombre scene. Animals clustered in thick knots to drink at sparsely filled water-holes, quickly trampling the moist earth into a quagmire. Others hugged the green gash marking the tortuous course of the Orangi River, a tributary of the Grumeti, where it cut a shallow path across the yellow plain. Impala mingled with a troop of olive baboons as stilt-legged giraffe moved elegantly past a truculent herd of buffalo. And everywhere I looked, long columns of wildebeest wended their way through the shade cast by thorn scrub, escaping from the dusty plains.

During the dry season, the Western Corridor provides a sanctuary for a large number of different animals. Among the commonest are the grazers – buffalo, zebra, wildebeest, topi and Thomson's gazelles. The fact that such a variety of species can co-exist is reflected in a complex range of physical and behavioural adaptations to their feeding environment. Many species avoid competition by occupying different parts of the habitat and by eating different types of grass. Even when consuming the same species of grass, the animals often select different portions with varying ratios of leaf to stem. Some also prefer certain stages of growth to others.

These differences in diet are reflected in the structure of the animals' mouths. Zebra, for instance, have razor-sharp teeth in both upper and lower jaws, enabling them to chop down the tall stems and seed-heads of the longer grasses. Wildebeest have broad muzzles with large, fleshy lips, ideally suited for mowing short leafy grass. Topi have long, aquiline features adapted for picking off green leaves buried among the taller stems. And Thomson's gazelles possess narrow, pointed muzzles to clip the newest shoots unavailable to their wider-mouthed relatives. The browsers are also separated according to their mouth anatomy and by the height at which they feed, ranging from giraffe down to the smallest antelopes such as klipspringer and dik-dik. Every conceivable niche is occupied by some creature or other – no opportunity is wasted.

The combination of grasses found in the Western Corridor reflects the impact of seasonal fires, which for centuries have swept through these areas during the dry season, thinning out the woodlands and enabling particular fire-adapted grass species to thrive. The commonest are red oat grass *Themeda triandra*, and the coarse bamboo grass *Pennisetum mezianum*. *Digitaria macroblephora* is also widespread and heavily grazed by the wildebeest. These pastures of long grass remain virtually ungrazed in the absence of the migratory herds. By the time the migration returns, the rainy season is over and some of the grasses have set their seed, standing more than a metre tall. The flowering stalks will soon become dry and fibrous, though still providing plenty of food for the zebra herds: zebra are non-ruminants, capable of digesting large quantities of coarse vegetation rapidly.

. . . to the smallest antelopes such as klipspringer . . .

Their progression through these long grass areas exposes the green leaf table preferred by the wildebeest, who eat less than the zebra but are able to digest their food more efficiently. Unless rain falls, little new growth occurs as the dry season closes in around the herds, forcing them to seek alternative pastures as the grass is whittled away by the effects of fire and grazing, trampling and natural decomposition.

In the months ahead, the animals concentrate at the rivers' edges where wild figs and podo trees grow alongside borassus palms, together providing a luxuriant touch of shade to temper the harsh land. Even in the driest years, the animals can always find permanent pools of water somewhere among the meanders and ox-bows of the Mbalageti and Grumeti Rivers. Meanwhile many of the zebra have already pushed northwards, leaving the Thomson's gazelles to occupy the woodlands at the edge of the plains. Their migration covers less ground than that of the wildebeest and zebra.

. . . and dik-dik . . .

I proceeded carefully, wary of arousing the wrath of a resting buffalo. But there was none to be seen. Figs and wild olives dotted the hill-top and there, amidst the silence, a pair of klipspringers perched like tiny statuettes atop boulders sculpted by the forces of heat and cold, wind and dust. Their large hair-filled ears flapped constantly as a blue and orange agama lizard darted back and forth to snap up the bothersome flies, at times pausing astride the back of the recumbent female while digesting his meal. The male klipspringer stood facing me, perfectly balanced on the blunt tips of his cylindrical hooves, the hard upright outer hoof-casing working against the spongy pad to provide purchase on the smooth rock walls. His coat was thick and coarse, the colour of salt and pepper. Each individual hair had a hollow, air-filled core helping to insulate the hardy little animal against the extremes of temperatures experienced high in the sky, and to protect against a bone-breaking fall.

I wondered for how many years these denizens of the craggy hill-tops had watched the mass exodus of wildebeest and zebra file past below them, while they stayed on, year round residents rooted to their own small territory.

All antelopes – there are more than seventy species in Africa – belong to the family *Bovidae*, the hollow-horned ruminants, which also includes the Cape buffalo. They are by far the most numerous and varied of all the world's hoofed animals, ranging in size from the shy dik-dik which stands about thirty centimetres tall, to the enormous cow-like eland that measures two metres at the shoulder and can weigh up to 700 kilogrammes. Collectively the antelope impart a Pleistocene aura to Africa's plains and woodlands. And nowhere is that more apparent than here in the Serengeti.

The word antelope is derived from the Greek *antholops* meaning 'brightness of eye'. Similarly, gazelles – which are antelope – derive their name from the Arabic word *ghazal*, meaning 'bright eyed'. The bulging eyes of klipspringers and wildebeest are placed strategically on the sides of their heads and are endowed with elongated pupils and highly sensitive retinas, enabling them to pick out the

slightest movement. But to survive amidst a wealth of predators antelopes need other specialisations besides good vision. They possess acute senses of smell and hearing, long legs and hooves for rapid flight, and a four-chambered ruminant stomach that allows them to remain vigilant while chewing and digesting their food.

Behaviourally and physically, the klipspringer and the wildebeest are about as far removed as any of Africa's antelope species: the one a small, sedentary, hill-dwelling species that feeds by browsing; the other a large, nomadic, grass-eating animal inhabiting the open plains. A pair of klipspringers spend their entire lives within just a few hectares while migratory wildebeest must roam over thousands of square kilometres of open plains and wooded grassland in search of food.

The klipspringer's lifestyle is thought to be fairly similar to that of its forest-dwelling ancestors, though the earliest antelopes probably lived on individual territories. Klipspringers, however, form monogamous pairs, and a male and female spend the majority of their time within a few metres of one another.

Territory is life itself to a klipspringer, providing sufficient food through drought and deluge, as well as ensuring a place to mate and raise a single young annually. Both male and female defend their territory against rivals and evict each offspring at maturity, forcing it to find a place of its own. When necessary they can go for months without drinking, obtaining moisture from their food.

Although all antelopes are herbivores, some are browsers like the klipspringer, which eats the flowers, fruits, seed-pods, young shoots and leaves of a wide range of shrubs and herbs. Others are grazers like the wildebeest, feeding on various species of grass. Some members of the family survive on a combination of both browse and grass, the proportion depending on the habitat and season. Impala exist in this manner.

As I turned away, the still air was shattered by a duet of trumpet-like whistles, repeated every few seconds. The klipspringers were signalling to me, and every other wild creature within earshot, that I had been seen – in effect telling me to move on, to abandon the hunt. They were acting just as they would on seeing a caracal or leopard stalking the rocky outcrops in search of prey.

Though all wildebeest are to a greater or lesser extent nomadic, not all migrate on the epic scale for which the Serengeti's herds are renowned. There are three separate resident populations within the Serengeti/Mara ecosystem whose movements and numbers are of a lesser order – numbering thousands rather than hundreds of thousands of animals. Two of these are found on the Ndabaka plains near Lake Victoria, in the Western Corridor; and in the Loliondo Controlled Area to the north east of the Serengeti. The third population, found in the Mara/Loita area of Kenya, migrates between wet and dry season ranges, preferring to give birth on the short grasses of the Loita plains. When the plains dry up, the animals must follow the rainfall gradient west if they are to find sufficient food and water, trekking to areas in and around the Masai Mara. Once

OVERLEAF *Not all migrate on the epic scale for which the the Serengeti's herds are renowned . . .*

the short rains begin in October or November, the Loita wildebeest turn to the east and trek back towards their own traditional calving grounds instead of journeying south with the Serengeti herds. During the dry season the Serengeti's migratory wildebeest mingle with these resident herds. Physically they all look the same, although each population breeds at slightly different times each year and so remains distinct from the others.

Wildebeest can only survive as a resident population if an area provides suitable grazing conditions year round. And that requires an adequate supply of rainfall. The Serengeti's short grass plains are unable to fulfil these requirements. From July through to October the grasses wither and water-holes remain dry. There is insufficient food and water for one wildebeest to survive, let alone tens of thousands. But suitable conditions are to be found in the areas of higher annual rainfall and wooded grasslands to north and west: the same areas that support the resident populations year round. Only by migrating can the majority of wildebeest bridge these seasonal and spatial variations in their food supply. Rapid movement from one area to another is their key to survival.

Wildebeest have long been characterised as ugly fools: the clowns of the plains; a random collection of parts left over from the creation of the other animals; God's joke. But in reality they are superbly adapted creatures endowed with the flexibility to survive in an unpredictable environment.

The migration we see today is a reflection of journeys travelled in times past: an event that has been happening for as long as wildebeest have roamed the Serengeti/Mara region. It is only the extent and timing of their wanderings that change. Their annual passage from the southern plains to the dry season holding areas is strongly directional, and under suitable conditions follows traditional, well defined routes. The movement between plains and woodland has certainly been in existence long enough for it to be genetically incorporated into their behaviour. But the availability of food within wet and dry season ranges is constantly changing. To survive, wildebeest must be capable of detecting suitable feeding sites on a day to day basis, and adapting their behaviour accordingly.

A remarkable feature of the wildebeest's nomadic wanderings is their ability to locate unerringly areas of good grazing, despite the fact that they may be many kilometres apart. It is never long before they arrive in the green areas. Ultimately it is rain that stimulates regrowth during the dry season, so it is thought that rain is the guiding light to their movements. Rain can be falling fifty kilometres away, yet the wildebeest somehow manage to be there in time to meet the surge of green shoots that showers quickly unleash and which the herds prefer. Do they hurry over the drying plains in response to the sight of lightning flickering across the darkening skies, or is it the sound of thunder that sends them galloping into the distance? Perhaps they can smell the rain with their large, sensitive noses, detect the dampness on the wind: it is probably a combination of all these senses that ensures they are in the right place on time.

But a wildebeest needs more than just a refined ability to locate its food. It

must learn the easiest way to reach such places, a lifelong process that begins on the short grass plains. A calf accompanies its mother throughout its first year, in effect receiving a guided tour of the current migration route. This provides a broad outline of the path to be followed in years to come: the location of preferred feeding sites, watering places and river crossings. Whatever instincts the calf is born with, enabling it to survive the rigours of migration, are broadened by individual experience and the natural tendency to follow others.

The only stable form of association the wildebeest knows in its life is the one forged at birth with its mother, a relationship that dissolves with the arrival of the next generation of calves. It seems unlikely that any animal could easily remain within the same herd from one season to the next. The fact that wildebeest do not form fixed relationships between individuals no doubt helps spread information more quickly through the population as to new and vacant areas providing good grazing, and the easiest routes leading to them. The herd is a storehouse of acquired knowledge which facilitates the flow of information from one animal to another.

. . . superbly adapted creatures endowed with the flexibility to survive in an unpredictable environment . . .

5 *The Fight to Save the Animals*

I go among the fields and catch a glimpse of a stoat or a fieldmouse peeping out of the withered grass – the creature hath a purpose and its eyes are bright with it. I go amongst the buildings of a city and I see a man hurrying along – to what? The creature has a purpose and his eyes are bright with it.

JOHN KEATS – LETTER, 1819

Tall and broad-shouldered, David Babu strikes an imposing figure. Some years after he graduated from his post of Chief Park Warden of the Serengeti to become Director of Tanzania National Parks, people still talk of him with respect. Nobody took liberties with Babu, and even today tour drivers from Kenya well remember his reputation as a tough warden who was intolerant of any breach of park regulations. It was he who implemented the rule forbidding vehicles from leaving the well-kept roads within sixteen kilometres of Seronera Lodge – a policy which reduced the impact of tourism where the pressure was greatest and dispersed it into more outlying areas. The rule persists to this day.

When I first visited Serengeti in 1975, we were fortunate in seeing both leopard and lion along the Seronera River. Naturally we were disappointed at not being able to drive closer to photograph them when they were located some distance from the track. But our ranger politely informed us that the continued well-being of the wild animals was the ultimate priority of a national park and that under no circumstance could we leave the road. Parks need more wardens like Babu.

No one is sure what effect car-loads of people have on wildlife, particularly on the predators which receive most attention. Does it actually affect their reproductive success or limit their ability to obtain sufficient food for themselves and their young? Most people who visit popular tourist destinations, such as Ngorongoro Crater or Masai Mara, are amazed by the apparent tameness of the wildlife. The animals barely appear to acknowledge the close proximity of vehicles: they seem content to graze or sleep, mate or kill, within metres of a rapt audience of visitors. Seeing this, it would be tempting to console oneself with the belief that all was well. But there is no doubt that the close attention of too many vehicles does affect the hunting success of most predators – particularly cheetah – forcing them to hunt at less advantageous times of the day or night, when competition from other predators is greater. Sometimes a female may be compelled to move young cubs from their hiding place simply to avoid undue disturbance, subjecting them in the process to the risk of predation. Regardless of these considerations, most visitors do not enjoy seeing a dozen or more vehicles crowded round a cheetah and her cubs, at times seeing them frightened. It offends our sensibilities and destroys any real sense of wilderness. It is unnecessary.

One day early in June 1986, while the wildebeest migration was passing through Seronera, David Babu visited the Serengeti to co-ordinate an anti-poaching raid into the Western Corridor. It was agreed that I could go along to see for myself the extent of the problem.

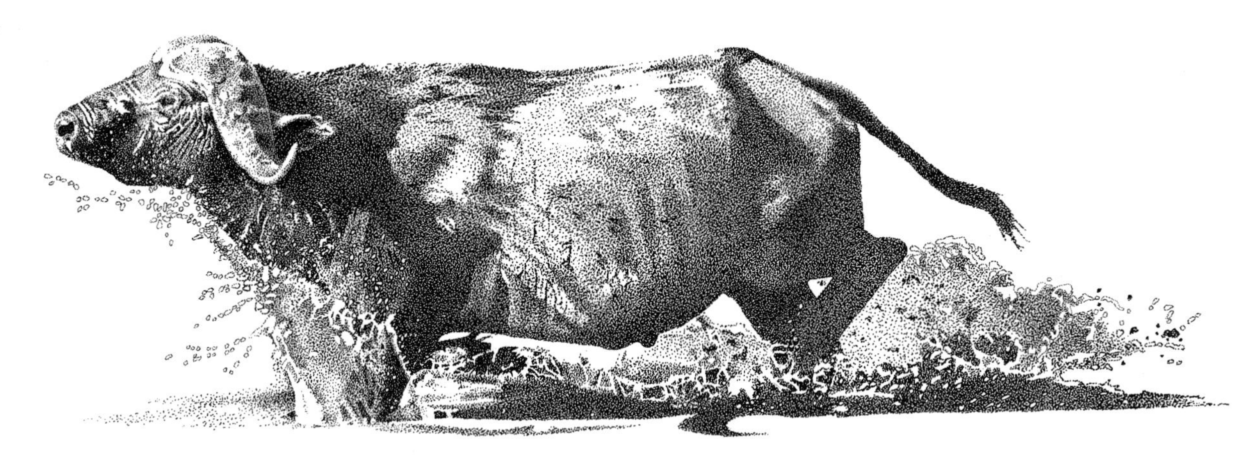

We arose long before dawn and drove to park headquarters where Babu was already briefing his army of rangers. Small groups of men dressed in battle fatigues were assigned to half a dozen vehicles, whilst drivers were instructed on the route to be followed. This was the first motorised sweep deep into the Western Corridor for many months, and every available vehicle had been commandeered from Parks and the Research Centre.

Tanzania is a big country. More than a quarter of its land surface is afforded some sort of protected status, with 8% set aside as national parks. Yet it lacks the resources to police such vast areas adequately. Vehicles, fuel and spare parts are prohibitively expensive, and as often as not unavailable.

Not surprisingly, there was an undisguised air of impatience in Babu's voice as he waited for the last land rover to arrive from the Research Centre. It had rained heavily during the night – the first rain in weeks. Babu knew that the track leading us through the Western Corridor would be waterlogged, delaying our rendezvous with the poaching gangs. It was essential that we moved quickly.

The column of vehicles turned off the main road and disappeared into the long grass and whistling thorn country of the western Serengeti. Lurching and skidding along the narrow pathway, the first vehicle soon began to falter. Its wheels spun wildly but before it had time to cut deeper into the boggy soil, men leapt from the backs of vehicles and quickly freed it from the mud. Babu remained tight-lipped as we drove onwards, visibly frustrated at being thwarted in the never-ending battle to protect the Serengeti. By the time the sky had brightened with the rising sun we were already too late for his liking.

Babu had spent some tough years as a young warden stationed at the Lamai guard post in the northern Serengeti, where he learned the ways of the poachers at first hand. The Lamai area has long been a hotbed of poaching activity, traditionally presided over by tough and aggressive Wakuria tribesmen. They made no secret of their contempt for the authorities, and considered it their right to hunt wild animals inside or outside the park. The poachers let it be known that they were quite prepared to kill anyone who stood in their way.

With settlements now pressed right up to the park boundary in some areas, the poachers could check their snare lines in the early hours and be safely back in the villages soon after first light. Our only hope of success was in surprising men grown accustomed to operating unhindered. It was vital that the anti-poaching unit made its presence felt now. For the past six months, the herds had remained protected on the central and eastern plains, many kilometres from the nearest village. But the next few months were critical as the migration pushed deep into the corridor and spilled over the park boundaries. The time of maximum returns for the poaching gangs was about to begin.

More rangers were waiting for us at Kirawira, an isolated guard post at the western edge of the Serengeti. The rangers greeted one another excitedly, pleased to see old friends after weeks spent in the isolation of the corridor. Everybody was anxious to get on with the job, to make contact with the enemy and to

OVERLEAF *It had rained heavily during the night . . .*

The rhino have almost gone . . .

relieve the frustrations of patrolling the surrounding area on foot, which proved of little consequence in such a vast area without more effective backup. The poachers definitely had the upper hand.

There are various categories of poaching afflicting the Serengeti. It is a disease from which some of the animals may never recover. The rhinos have already gone, sacrificed for the price of their horn. Twenty years ago in Africa there were 65,000 black rhino. Today there are less than five thousand. I saw no sign of these ancient beasts while living in the Serengeti, and it seems doubtful that more than a handful could have survived the well-organised onslaught of the commercial poaching gangs that are intent on bleeding the country dry of its animal wealth. The gangs still greedily roam the park for other lucrative trophies and are now concentrating their murderous efforts on the remains of the Serengeti's beleaguered elephant population.

There was a time when four thousand elephants roamed the Serengeti's woodlands. Now fewer than five hundred survive. It is a very uneven contest: a handful of ill-equipped rangers struggling to contain the onslaught of poachers armed with automatic weapons and supported by vehicles. Today's sweep was in search of meat poachers who indiscriminately kill wildebeest, zebra, topi, hartebeest, buffalo, impala, giraffe and gazelle. Wire snares, poisoned arrows, pitfalls and spears are their weapons, just as they were thirty years ago.

Nobody really knows the true size of the poaching problem. In the late 50s, Colonel P. G. Molloy, M.C., director of Tanzania National Parks, estimated that no less than 150,000 game animals were being poached every year in and around the Serengeti. Without in any way wishing to play down the enormity of the problem, this figure appears to have been an exaggeration. Currently research workers estimate that 40–50,000 wildebeest are killed annually but nobody doubts that it may be more. And in the north and west, the buffalo population has dwindled to less than half 1970 levels. So even more wildebeest and zebra may be killed in the future. Suffice it to say that game meat from Serengeti is now distributed as far north as Kisumu in neighbouring Kenya.

It was decided to split the vehicles up into three separate groups so as to maximise the chances of surprising the poachers. Before long we came across the first signs of human activity inside the park, a wire snare carefully tied into position between bushes across a game trail. We continued along the edge of the plains, following the line of the thorn scrub and circling likely looking thickets. But though we searched carefully we could find no further evidence of the poachers.

Suddenly, one of the rangers signalled the driver to slow down. I strained my eyes in the direction in which he was pointing, yet could see nothing untoward about the tangle of roots and branches comprising the dense thicket of cover beyond the plain. It was the kind of place where a lioness might seek shelter from the midday heat, or safely leave her small cubs while she went off to hunt: the same place that human predators frequent when surprised in open country.

*The buffalo population has dwindled . . .
Thousands of wildebeest
are killed annually . . .*

But the rangers were wise to the ways of the poachers and could spot a bare foot or tattered shirt among the confusion of leaves and shadows. Long before the vehicle had rolled to a stop, a handful of armed men were racing towards the thicket, shouting an intimidating mixture of threats and at the same time jabbing the barrels of their guns through the undergrowth.

A small group of rag-tag men emerged from the darkness, their garments so torn and threadbare that they could hardly be described as clothes. The rangers prodded and booted their captives, anxious for information as to the whereabouts of the poachers' camp: how many accomplices? How far? Were there guns? There was no time for niceties. This was a war.

Shackled to one of his captors by a piece of rope tied around his waist, the youngest poacher was frog-marched into the bush, while his accomplices were herded into a patrol vehicle. We continued cautiously through the long grass in the footsteps of the prisoner – avoiding every dry stick and loose rock so as not to alert our quarry.

The poachers' hide-out was situated in a thick patch of bush bordering a stream bed. As we approached, one man broke from cover. But the rangers were after him in a flash, running him down and wrestling him to the ground. Strips of fresh meat lay everywhere, spread in circles across a table of grass to dry in the sun. Elsewhere, concealed from the eagle-eyed vultures, whole portions of wildebeest had been carried into the trees to await butchery, secured high above the ground beyond the reach of lions and hyaenas – just like a leopard's kills. Spears and piles of wire snares rested against trees; bows and quivers full of arrows hung from branches within the dark interior of the hunters' lair. It was obvious that the last thing the poachers had expected was a visit from The Law.

I am not quite sure what I had imagined when we set out from park headquarters – what I expected to see, or feel. On one level I relished the excitement of the chase as we searched for these men. I had visualised the poachers as criminals, tough and defiant. Instead they just looked poor. These were not the middlemen – the fat cats who live off the proceeds of commercial poaching for meat and trophies, and grow rich. Those men sit shielded behind an impenetrable web of corruption which virtually ensures their freedom. It is rarely the big men that languish behind bars. Yet the sheer enormity of the poaching problem leaves little room for this sort of distinction.

In the old days, the poachers used snares fashioned from wild sisal, and dug long lines of deep game pits across the time-honoured paths of the migratory animals. Their arrows were flighted with vulture feathers and coated with a deadly poison brewed from the bark and twigs of the *Acokanthera* plant – they still are. These were the traditional hunters who killed to feed themselves and their families, as innocent in their lifestyle as the bison-killing Plains Indians of North America, or the Bushmen of the Kalahari. They were people enacting a way of life passed down from father to son, from one generation to the next, who took pride in their bush craft and in fashioning the tools of their profession.

They killed to survive.

Since World War II, the inefficient rope snares have been replaced by lengths of steel cable which never deteriorate and can be used time and time again. The loops of wire are held open by strands of sansevieria fibre attached to twigs and anchored to a tree or log. These snares are hideous devices set along game trails, or near to water-holes: they tether their victims by neck or foot, biting bone-deep into an animal's flesh. Occasionally the struggling captive jerks itself free, then wanders for days with its leash of steel wire firmly embedded in a festering wound. During Myles Turner's eighteen years in the Serengeti, the anti-poaching force destroyed 22,000 of these deadly wire nooses. Imagine how many dead animals they had accounted for over the years.

Gangs of up to twenty men, sometimes armed with hundreds of snares, ply their cruel trade to good advantage each year as the migration passes through the acacia thickets and commiphora woodlands. Dozens of animals may be trapped in a single day in snares spaced out over hundreds of metres of bush country. To prevent the vultures from spoiling the carcass, the poachers slash a snared animal's legs, severing the hamstrings to keep it alive until it is ready for butchering.

With large numbers of people clustered along the western boundaries of the Serengeti, there is a ready market for the meat poachers; in many places it is a highly commercial operation and has been for years. The poacher no longer hunts simply to feed his family. He has become part of the cash economy. He is the poor man's butcher, tapping a rich source of free meat. The more he kills, the more he is paid. He is not nomadic, he stays in well-established settlements at the very edge of the park, and his most deadly and effective weapon is still the wire snare.

I expressed my ambivalent feelings about the meat poachers to Babu as we drove back to Seronera. But he was both adamant and convincing in his reply. People had fought long and hard to secure the boundaries of the Serengeti. Men had died in trying to fulfil the pledge to save the animals. It was unlikely that the Serengeti would ever be enlarged. The rapidly increasing human population would see to that. The priority was to protect the herds for as long as they remained within the park boundaries. The killing of wild animals in parks and reserves was illegal. Nothing else mattered.

Any sympathy I might have harboured for these men was tempered later by the knowledge of an incident at one of the guard posts. A gang of armed poachers attacked during the night, firing indiscriminately through the fragile walls of the rangers' houses, killing two rangers before making off into the darkness. Somehow poaching has to be controlled. But who is going to provide the resources? The outside world must be prepared to play its part.

Before the devastating arrival of rinderpest at the end of last century, the migration was probably on a scale similar to that of today, confined only by ancient physical barriers: the Isuria Escarpment to the north west, forested hills of Loita and Gol in the east, the Crater Highlands and Eyasi Escarpment along the south, and Lake Victoria to the west. The size of the human population would have had little impact on their wanderings.

During the rinderpest years, the wildebeest population is thought to have been no larger than 250,000 – one fifth of today's huge army of animals. In those days the migration had no need to travel north as far as the Mara Reserve. But as the population swelled, the herds began to leave the plains earlier – particularly in drier years – and were forced to venture further north in their quest for food and water.

Nowadays, many of the Serengeti's wildebeest do not enter the Western Corridor at all after leaving the plains. Instead they travel directly north as the dry season sets in, tarrying a while in the Loliondo Controlled Area to the north east of the park after leaving the Serengeti plains. Meanwhile those animals that departed for the Western Corridor at the end of May eventually exhaust the best of the grazing in the valley pastures. Then they must move further north and enter the areas known as the Lamai Wedge and the Mara Triangle.

Last year, more than a million wildebeest advanced across the border into the Masai Mara in Kenya. The Mara and Lamai Wedge now provide vital dry season pastures for the hungry herds in the northern part of their range. Without these areas of higher rainfall to sustain them, the migratory populations would soon collapse and the effect would reverberate throughout the whole plant and animal community comprising the Serengeti/Mara ecosystem.

6 *Mara – the Spotted Land*

There are two Africas and I do not know which I love the best: the green, lush, bright country when the sap is running and the earth is wet; or the dry, brown-gold wastes of the drought, when the sky closes down, hazy and smoke-dimmed, and the sun is copper-coloured and distorted.

DORIS LESSING – GOING HOME

Where they meet there is nothing to distinguish the Mara from the Serengeti. The distinctions are arbitrary. They are one land, joined by the continuity of river and scrubland, green hills and clear blue skies: undivided by fences which would spell disaster for the migratory animals.

The Mara may lack the breath-taking open spaces of the southern Serengeti. But few would dispute that it deserves its status as one of the world's finest game-viewing areas: it is not uncommon to see all of the 'big five' in a single day – elephant, rhino, buffalo, lion and leopard: the yardstick by which most tour guides measure a successful safari. Where else can you watch a pack of wild dogs hunting to feed their eighteen puppies, and then, as the sun dips below the Isuria Escarpment, sit spellbound as thousands of wildebeest brave the crocodile-infested waters of the Mara River?

Ironically, the closure of the Kenya/Tanzania border in 1977 has had much to do with the dramatic development of the Mara as Kenya's foremost game-viewing area. In the early 70s, the northern Mara was still scarcely visited: it was a haven for a handful of local tourists and visitors to Governor's, an exclusive tented camp set in the heart of the best game-viewing enclave in Kenya; a chance for those familiar with the Mara to escape the more crowded safari destinations such as Amboseli National Park. That level of tourist trade had little impact on the area.

In those days, the Mara was synonymous with Keekorok Lodge, situated in the southern part of the reserve. The Nairobi-based tourist industry, catering for travel throughout East Africa, viewed Mara as little more than an overnight stop at the end of the 'milk run' through the parks of northern Tanzania. Whatever the Mara packed into its well-rounded 1,510 square kilometres had been seen at least once already: elephants and tree-climbing lions in Lake Manyara; rhinos and black-maned lions in the Ngorongoro Crater; wild dogs, cheetahs, leopards and more lions in the Serengeti. No matter then that the Mara could boast all of these within its boundaries. When the tour buses bumped their way back across the border, bones were weary from the jolt and judder of rough roads and sagging beds. Thoughts were already turned towards home – a safari well spent, an end to dawn rises. One final stop to glimpse a pink paradise called Lake Nakuru, where more than a million flamingoes sometimes reside, before last minute shopping in Nairobi for curios and carvings. Some visitors hardly even noticed the Mara in those days – what was the name of that place after Serengeti?

This was the way things were until February 1977. Then Tanzania closed its border on Kenya's tour operators. Kenya responded quickly by developing and expanding tourism to its own wildlife areas. Soon it was the Mara that everyone wanted to visit.

Officially the border between Kenya and Tanzania re-opened in October 1983. But the Tanzanians had already determined that Kenya's tour operators would no longer be permitted to profit unfairly from their spectacular wildlife areas. People wishing to visit Tanzania would do so on their terms. Admittedly, things might not run quite so smoothly for a while (and indeed in the early stages there were plenty of inconveniences for tourists when measured against the sophistication of Kenya's travel industry). But that was the price visitors would have to pay to gain access to some of the world's greatest natural treasures: Lake Manyara, Tarangire, Ngorongoro Crater, the Serengeti. And for the really adventurous, there was the chance to take a tented safari to those vast wilderness countries to the south – the Selous and Ruaha.

Today, the border posts at Sand River and Bologonja, providing direct access between Mara and Serengeti, remain closed to the tourist industry, though Kenya residents are entitled to travel freely by this route. Tour operators must either arrange for their overseas visitors to be collected at Namanga, the main port of entry into Tanzania, or fly them directly to Arusha by air charter. In future, travel within Tanzania will be in Tanzanian registered vehicles, driven by local driver-guides. Couriers from Kenya must now apply for work permits if they wish to accompany their own clients. The days of the milk run are definitely over.

It is 6.30 a.m. and a tour guide is explaining to his group that after a night of heavy rain, it is advisable to have breakfast first in the hope that the sun will emerge to dry out the roads; that to go out on an early game drive will only mean getting stuck and everyone will have to get out and help push. And that means getting dirty. The visitors are understandably disappointed; one elderly gentleman mutters to no one in particular that he is sure that the brochure said the long rains should be over by now. And where is this migration that everyone told him about? They decide to go out anyway, despite further words of caution from their courier. The previous evening another group of tourists had regaled them with tales of the elusive leopard and a pack of wild dogs. Why miss the opportunity of seeing such rarities for themselves? Their guide admits defeat, smiles knowingly and ushers them into the waiting minibuses. Half an hour later, all three vehicles are hopelessly bogged down in the Mara's heavy black cotton soil.

Everyone gets stuck in the Mara during the rains. Yet I long for their arrival, for they herald the onset of the low season. By October the ground looks exhausted – ravaged by the wrenching of hungry mouths and the tread of a million hooves – as if it will never recover. But soon the land sparkles in a new coat of glittering green. During January and February the skies brighten and the land becomes dry again. Then from March to May the long rains feed the lush

grasses as a wind fuels a fire, until they sprout waist high, and the spiky oatheads clog the radiator of my vehicle.

At this time, the Mara seems strangely empty, though in fact the abundant resident game is sufficient to stun the senses of visitors used to exulting at the sight of an occasional coyote, badger, deer or kestrel. Topi stand watch atop termite castles, staring out across the grasslands as sombre herds of buffalo reclaim the vacant plains and drop their calves; giraffes crane their elegant necks to browse the taller trees, while warthogs struggle to see over the tops of the long grass.

Soon the land takes on the ominous appearance of yet another of the giant wheat-fields that are pushing ever closer towards the Mara from the east. Access to some areas becomes impossible – sometimes for months – making for lengthy detours to circumvent the water-logged terrain. Musiara Marsh once more reverts to an elephant's paradise, where herds a hundred strong slop through belly-high sedges to feed and wallow. And woe betide the vehicle that tries to follow. The well-worn tracks providing easy access to the very heart of such areas in the dry season can imprison you for a day at a time during the rains.

The short rains are usually expected from the middle of October through until December, but in 1986 they were late. The Mara received little rain until the end of November. Then throughout the first half of 1987 it experienced one of the wettest periods in recent memory. In fact it rained fairly heavily from January till June: now, in the first week of June, the grass testifies to that. Yet perhaps the Mara is at its most beautiful during the rains: everywhere you look you see long green grass, distant blue hills and to the north the confining beauty of the Isuria Escarpment.

As June departs, the Mara resembles a sea of grass which laps over the side of the heavy metal guard protecting the front of my vehicle. All this morning I have driven through kilometre after kilometre of seemingly empty plains. Many of the familiar dry season tracks are obliterated from view: the land looks different. There is an air of expectancy in the gentle breeze that rustles the drooping oatheads, and I wonder how far I will have to drive before I find what I am searching for – thousands of wildebeest marching through the Mara on their annual pilgrimage in search of food and water.

The sight of those first black specks, spilling like a dark stain across the gently undulating landscape, or plodding one after another in long thread-like columns, never fails to send a wave of excitement through me. For each of the last ten years I have awaited word of their arrival. My mind races ahead – the frantic sounds of a river crossing, the breathless silence of lions crouched in ambush, the joy of watching wild dogs at their den site near Aitong. It is like the beginning of a new year.

Time in the Mara is measured by the arrival and departure of the migration: their appearance breathes life into the tall grasslands, their departure leaves the

OVERLEAF . . . *those first black specks spilling like a dark stain across the gently undulating landscape* . . .

Tiny malachite kingfishers wait for the chance to catch a fish . . .

earth exhausted. By June, tour drivers and safari guides can be heard anxiously enquiring about the whereabouts of the wildebeest, and pilots ferrying tourists to the camps and lodges are besieged with questions from land-based residents as to the movements of the herds.

Everyone longs for the moment when the wildebeest return to mow down all that grass, so that once more the Mara stands unrivalled as *the* place to see lions, leopards and cheetahs. But regardless of when they cross the border it still may be more than a month before the migration emerges from over Rhino Ridge and is visible from a breakfast table at Governor's Camp. In 1984 it was the 13th August, in 1985 the 4th August and in 1986 the 4th July.

At last they are here, sweeping in a great wave across the Sand River, the intermittent water-course that separates the Mara from the Serengeti. By early July, the animals have pushed north west past Ol Kiombo and settled around the edges of Musiara Marsh. Here the nomadic herds are assured of water year round, regardless of season. A perennial spring feeds the kilometre-and-half-long marsh, creating a never-ending stream of clear, sweet water that flows from the base of a low escarpment, dividing the lush grasses surrounding the marsh from the scattered acacia thickets to the north. This is a bird-watcher's paradise, where fish eagles, saddlebill storks and tiny malachite kingfishers all wait for the chance to catch a fish.

Each evening the wildebeest herds gather on the surrounding plains and march in single file to the edge of the marsh. The animals drink quickly as the light fades from the sky, then hurry back out on to the plains or move to higher ground, bunching together for safety, distancing themselves from the dense reed-beds. Next morning the rhythmic booming call of ground hornbills greets a new African dawn, and the blood-red sun casts its glow across the dew-laden marsh. The herds return – or perhaps they are different herds, for by now the marsh is alive with wildebeest and zebra – galloping down from the higher ground, grunting and braying, the land reverberating with the sounds of life.

In places the blanket of animals is broken, where a dozen hyaenas feast on the remains of a carcass. Not far off is another area devoid of game. Here a pride of lions, their breath raw and steamy in the chill morning air, growl and squabble over a wildebeest and its five-month-old calf which they ambushed during the night.

When I first came to live in the Mara, it was to the marsh that I drove each morning. My guide and mentor was Joseph Rotich, the head driver at Mara River Camp. Through his eyes I came to learn something of the Mara – where to look for lions and leopards, when to cross the treacherous black cotton soils and where to go when you could not. In those days, the marsh was ruled by a pride of lions that I called the Marsh Lions. The name was not meant to imply that there was anything different about these particular lions: only that over the years

The animals have settled around the edges of Musiara Marsh . . .

I came to know them as individuals. They were the lions of Musiara Marsh and before long I could recognise each pride member. Gradually, with Joseph's help, I learned where else to look for them when they were not to be found lurking around the fringes of the marsh.

During the dry season, when the migration of wildebeest and zebra flooded their territory, the Marsh Lions confined their hunting activities almost exclusively to within a few square kilometres of the marsh. Between July and October it was the core of their pride territory, a jealously guarded prize at the heart of their domain, defended fiercely from the incursions of neighbouring prides. Three kilometres away to the east roamed the Miti Mbili pride: a group of four lionesses and their half-grown cubs left unprotected by pride males since the arrival of a pair of aggressive nomadic lions. To the south, beyond Rhino Ridge, lived the Paradise lions, the largest of all the prides in the northern Mara – a group of six adult lionesses, their numerous offspring and three fierce ginger-maned males. These were the buffalo killers who regularly overpowered full-grown Cape buffalo with the confidence born of years of experience at this most dangerous form of predation. North of the marsh, a shy group of lions named the Gorge pride roamed the acacia thickets and rocky outcrops of Fig Tree Ridge and Leopard Gorge where Joseph took me to search for leopards. Theirs was a

I learned where to look for the lions . . .

. . . beyond Rhino Ridge lived the Paradise lions . . .

small pride, an alliance of only a few lions forced to take their chances amongst the Masai tribesmen rather than venture far into Marsh Lion territory – where Brando, Scar and Mkubwa would be waiting to defend the land they shared with five lionesses: the Talek Twins and the Marsh Sisters. And to the west, past the marsh and across the Mara River, lay the country controlled by the Kichwa Tembo pride, home of Old Man, the one-eyed male who had been chased from the Miti Mbili pride, and who had finally found refuge on the west side of the river.

It was during my first game drive in the northern Mara that Joseph introduced me to Old Man. He had spotted the two males from several kilometres away. My heart missed a beat when he said, 'Lions – do you want to have a look at them?'

Did I want to have a look at them? I wanted to look at *every* lion I could find.

 Nowhere was a finer-maned lion to be found . . .

Old Man lay with his companion Dark Mane in the shade cast by a solitary tree at the edge of the marsh. Joseph explained to me that these two males, together with a younger lion called Mkubwa, dominated this part of the Mara. They were the largest lions I had ever seen, each crowned with a luxuriant mane of hair. I can still vividly remember the excitement I felt on looking at those lions, relishing the thought of being able to learn more about them in the years ahead, in knowing that the Mara was to be my home. Joseph looked at me and grinned. Not a word was spoken; he understood.

In those days, Old Man was known simply as the Blond Male in recognition of his fine mane, coloured the pale tones of dry season grass. It was only years later – long after Dark Mane had vanished – that the Blond Male became known as Old Man. As he aged, his mane turned a deep ginger, his chest hair as black as coal. By that time his left eye had been gouged out in a fight over territory, and he had transferred his allegiance to the Kichwa Tembo pride in the Mara Triangle. People used to travel many kilometres in the hope of catching a glimpse of the old warrior – one-eyed and grizzly-faced – for nowhere was a finer-maned lion to be found. His story taught me something of the life of a pride male – a tough life, with many injuries and testing battles over territory to be endured.

All male lions are forced to leave the pride in which they are born when they are approximately two and a half to three years old. This is encouraged by the increasing hostility of their older relatives and helps prevent inbreeding. The females within a pride are related to each other – grandmothers and mothers, sisters and daughters, aunts and nieces forming the stable core of each pride. Lionesses live until they are about fifteen years old, which means that at times there is a surplus of young females who must leave the pride and survive as best they can. Surprisingly, these females do sometimes manage to establish themselves as a separate pride in a new area.

When not resident in a pride, males must lead the life of nomads. These roaming bands of males are often brothers or cousins who have remained together since leaving their natal pride. But sometimes during their nomadic years they form an alliance with unrelated males. These coalitions are a constant threat to pride males. Each time a challenge is successful, the new males try and kill any young cubs and chase away the sub-adults: a necessary prerequisite for reproductive success, as cubs are dependent on their mothers for the first one and a half to two years, and during this period the lionesses remain infertile. A few months after the death of their cubs the females are able to conceive the cubs of the new pride males. The larger alliances of males are better able to defend a territory and hold tenure for longer, thereby helping to ensure the stability necessary for their own cubs to survive.

The majority of Mara prides are lorded over by two or three males, though larger groups of as many as six or seven sometimes appear. But, with age, even the strongest alliances fade and are in turn deposed by younger or numerically stronger groups of males.

Old Man is long since dead, one of the most successful males ever to roam the northern Mara. By the time he disappeared he had associated with at least four different groups of lions – the Miti Mbili, Gorge, Marsh and Kichwa Tembo prides – each of which raised some of his cubs to maturity. He was at least twelve years old when I last saw him. Joseph thought he might even be fourteen – old indeed by male lion standards. He lived out the final years of his life on the plains surrounding Kichwa Tembo. I often visited him there, and each year he appeared a little stiffer, a little slower – but still fit enough to mate. Some of his daughters reside in the area to this day: big, blonde lionesses with cubs of their own.

It was during the migration of 1982 that Half-Tail and his two companions first appeared around Kichwa Tembo. Their arrival came as no surprise. Each year when the wildebeest move into the Mara there is a noticeable increase in the number of nomads passing through the pride territories. These footloose males roam over huge areas: scavenging on the body of a wildebeest that has died of starvation, robbing hyaenas of a calf they have killed or warring with other lions for possession of a zebra carcass. Sometimes they kill for themselves. Many of the nomads are in their fourth or fifth year, powerful, hungry creatures: hungry for food and hungry for a territory to call their own. Some will have travelled from far away places in the Serengeti; others are exiles from Mara prides. The moveable feast provided by the migration frees these males to search far and wide in their quest for territory. Consequently the arrival of the wildebeest and zebra signalled a particularly dangerous time for Mara pride males, and it was not unusual to see a new era unfolding by the time the herds departed in October.

The three new arrivals were not alone in coveting the Kichwa Tembo territory and its lionesses. Two other groups of males had also arrived in the area. Throughout the dry season, these alliances sought to gain control, to intimidate the others into leaving. Despite their size and impressive looks, the Kichwa Tembo males – Old Man, Scar and Mkubwa – behaved increasingly nervously, acting like prisoners within their own domain. For a while, nothing untoward happened. Then Mkubwa disappeared.

A few days later I chanced upon Old Man lying under a tree far to the south of the Kichwa Tembo plains. He was alone. As I drove towards him, he sat up – unusual for the old lion: normally he ignored the sound of an approaching vehicle. I searched for signs of injury. But none were to be seen: no bullet-shaped holes inflicted by the dagger-like canines of a rival male. Old Man kept looking around, nervously staring in the direction of those familiar plains and luggas that he had helped rule for the last three years. I knew then that I would probably never see him again. And I was right. It was only a few weeks later that Mkubwa was found by drivers on the other side of the river. He was pathetically thin, his body covered with deep bite marks. Later that day he died.

So Old Man and his companions had been forced to capitulate, and three scruffy nomads had come of age. For the moment they had prevailed, though it might be months before stability returned to the plains and the lionesses bore cubs for the new males. Night and morning they roared their challenge across the plains, staking their claim to the land. But the only answer came from far away, beyond the river: the grunting of their new neighbours, the Marsh Lions.

Pride territories vary in size, reflecting differences in habitat and availability of food. Some Mara prides occupy areas as little as twelve square kilometres, others require more than a hundred. The occupants spray-mark and roar to declare their presence, but an area of this size cannot be exclusively defended. Territories often overlap at their edges, and may shift in relation to one another from one season to the next.

The Kichwa Tembo, Miti Mbili and Paradise prides still patrol the same territories they occupied eleven years ago. Other territories have dwindled over the years, just as some have expanded, reflecting the changing fates and fortunes of the various prides. The Marsh Lions' territory is now divided, due to the exiling from the pride of three young lionesses who later took control of the hunting ground on the west side of the marsh. Though related to all the other females in the Marsh pride, these three lionesses now act as a separate pride, even though the three pride males control the entire area and mate with both groups of females.

All the other prides at times encroached into Marsh Lion territory, though they rarely penetrated as far as the marsh itself. On one occasion when this did happen, a major confrontation took place when members of the Miti Mbili pride and their cubs rested in the shade of a tree at the northern edge of the marsh. After a brief but bloody encounter, the five females from the Marsh pride routed the Miti Mbili lionesses, chasing them back into their own pride's territory.

Much of interest occurred after dark, the time when lions are most active and visitors are safely back in camp sipping sundowners around the camp fire. But it was often still possible to piece together events that had occurred unwitnessed – from the accounts of other drivers, the remains of a kill, the sounds of battle, injured lions and the ever-changing location of members of the different prides.

During the dry season, I would sometimes find lions from the Kichwa Tembo pride camped for days along the western banks of the Mara River, the edge of their territory. In the early days they emerged into daylight from the concealing darkness of the riverine forest. Now it is possible to see clearly through these previously dense stands of ancient trees – figs, wild olives, and African greenhearts that once hid from view the beauty of the Kichwa Tembo plains and the Isuria Escarpment beyond.

When no river crossings were taking place, the Kichwa lions would stand and stare keenly across the river at the rich hunting grounds around Musiara Marsh where daily during August and September they could see and hear the herds congregating to feed and drink. At times they would watch the dense throng of animals explode from the reed-beds, lost for a moment in a cloud of dust as the herds thundered away from the place where mud-spattered lionesses struggled to overpower a wildebeest that had paid the ultimate price for incaution.

And sometimes, when the Marsh Lions were elsewhere, the Kichwa Tembo pride would take their chances and wade up to their necks through the murky waters of the Mara River to lie in ambush themselves. Having killed, they would feed quickly, looking nervously about them for signs of their enemies, showing all the anxious signs of trespassers. Within minutes their secret would be revealed as vultures plummetted from the blue sky, indicating the presence of yet another kill in the marsh. Before long hyaenas dragged themselves from their wallows to track the vultures' line of descent. And far off to the south, a lion might look up and see the circle of dark specks spiralling down to earth. First one lion, then another, would get to its feet, stretch and yawn, before plodding off in the direction of Musiara Marsh. But by the time the Marsh Lions arrived to sniff and spray around the kill site of these alien lions, the Kichwa Tembo pride would have retreated back across the river into their own pride territory, distancing themselves from the harsh roars of their enemies.

For days now the wildebeest had massed on the rolling plains to the east of Musiara Marsh, trekking back and forth to places providing water: the murram pit next to Governor's Camp airstrip and the marsh itself. In recent times the Marsh Lions have concentrated their activities around Miti Mbili Lugga, a favourite place for the lionesses to raise young cubs. Each evening as the sun drops beneath the escarpment the lions emerge from the grass-choked lugga to stare out across the plains at the swelling herds.

Miti Mbili – the two trees – for years provided a visual landmark for drivers discussing the whereabouts of lions, cheetahs and rhinos. 'Two big males just south of Miti Mbili' or 'Head straight for Miti Mbili and you are bound to see the rhino.' Now the ancient trees have fallen, no longer a roost for vultures or a larder for leopards. Yet death has not robbed them of a function. Hyaenas and warthogs rest in their shadows; harrier hawks flap and scuttle over their dry limbs, poking a long scaly leg deep inside a hollow to claw out a giant beetle or lizard. And lion cubs sometimes play on the trellis of dead branches, as gradually the insects mine the store of trapped nutrients and return them to the soil.

The death of a landmark tree may seem like the end of an era to the human observer. Yet change is an integral part of all ecosystems. During the last hundred years, the Serengeti/Mara has seen dramatic variations in the balance between grass and woodland. Diseases such as rinderpest; the impact of tsetse flies on livestock in the area; rainfall, fire and elephants have all played their part. In consequence certain areas of the Serengeti/Mara have changed from open grassland to dense woodland and back in less than a century. Year by year the tree-spotted landscape within the Mara Reserve is reverting to open plains and the luggas are gradually being stripped of trees and bushes.

. . . harrier hawks poking a long scaly leg deep inside a hollow . . .

. . . the burgeoning elephant population . . .

In recent times, the wildebeest have spent more than four months of each year in the Mara. By the end of the dry season, the animals have eaten all the grass. Consequently, fires are now a rarity on the plains. There is evidence that the sheer volume of wildebeest has become an important factor in preventing the resurgence of acacia trees and croton thickets. When wildebeest move through an area, they crop the grass to ground level. In the process, thousands of tiny seedlings, immersed among the tall stands of grass, are indiscriminately bitten and trampled as the animals harvest the pastures. And the bulls horn and batter the bushes, tearing the bark and breaking the branches. Elephants deftly grasp the seedlings with outstretched trunks, then gently nudge them free with a huge front foot. Giraffe and impala, gazelles and other small browsers all contribute to the effect. By the time the seedlings sprout anew, the wildebeest are back again.

At the south end of Miti Mbili lugga the vegetation still grows thickly. But each year the burgeoning elephant population strips the foliage, tearing off the lower branches and snapping the brittle bones of the trees with a sound like gunfire. Here a cul-de-sac of trees and bushes encloses a flat expanse of bare ground known to drivers as Bila Shaka – 'without doubt'.

Without doubt, it is here that the Marsh Lions and their cubs can often be found. The lions at Bila Shaka are invariably marked for kilometres around by a cluster of green Toyota Landcruisers. The drivers from the various camps monitor the activities of the predators from one game drive to the next, keeping watch over them as carefully as any scientist. Even when the grass grows higher than a stalking lion, they can usually find at least some of the pride members.

The location of favoured places such as Bila Shaka and the ancient fig trees at the edge of the marsh are passed on from one generation of lions to another. Through regular use they become familiar to mothers and daughters, granddaughters and nieces, significant to the lions as fruitful ambush sites or safe places for raising cubs. It is not uncommon for a lioness to give birth to cubs in the area where she herself was born.

. . . standing immobile in the grip of a lioness . . .

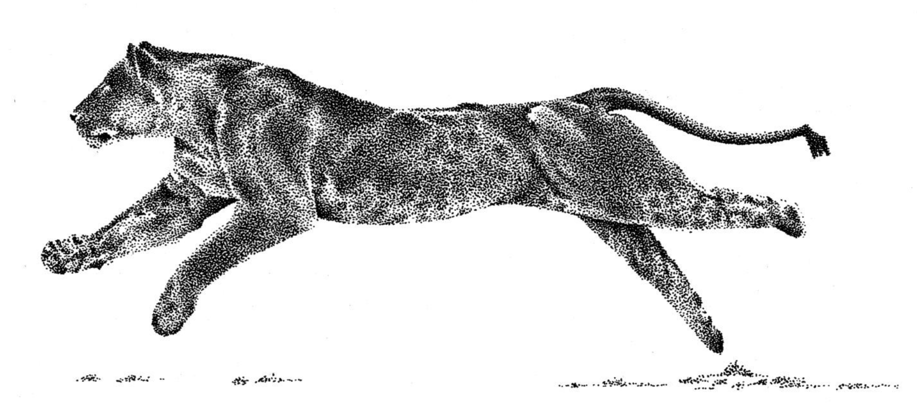

When the migration camps within the vicinity of the Marsh Lions' territory, there are days when they make multiple kills. The lionesses spread out like five tawny fingers in the long grass until they have virtually surrounded part of a herd. But lions do not seem to have learnt the folly of stalking their prey from up-wind and sometimes the wildebeest catch the scent of the predators. Then they bunch together, ears pricked, dark eyes bulging to catch sight of a stealthy movement in the grass, rushing this way and that in their panic to escape from the owners of that rank predatory odour.

Dusk is an especially vulnerable time for the herds. The advantages of living on the open plains quickly diminish, as the darkness favours the specialised nocturnal eyes of the hunters. If the wildebeest try to break free of the stranglehold of advancing lions, the cats will attack anyway. If they stand their ground, jostling and bumping against each other, the lions creep ever closer, making sure of success. Suddenly a lioness runs low through the grass. Without seeming to break stride, she is galloping, eating up the ground between her and the herd. Too late, they turn and try to flee. Blundering away through the singing grass, a dark shape suddenly falters, smothered with tawny fur, feeling ten sharp claws sink into its rump before standing immobile in the iron grip of a hundred and twenty kilogramme lioness.

There are days when the lions pick off three or four animals from the passing herds. I once counted half a dozen fresh carcasses in a single morning on Paradise plain within a radius of a few hundred metres. The lions lay among the dead, bloated with meat. Yet they refused to seek shade, determined to deny the vultures their feast. Rarely did I find them feeding on a single kill at this time of year.

The number of lions living within each pride is determined by their ability to survive during the lean months when prey is scarcest, not by the seemingly inexhaustible offerings of wildebeest and zebra. For as long as the wildebeest continue to blunder through the long grass, and pile up in their thousands to cross the Mara River, the lions prosper. There is sufficient food to support all pride members. Half-grown lions can refine their hunting skills: even if they stalk and are seen, chase without killing, wrestle a wildebeest to the ground and then lose it, they will not starve as a consequence of their inexperience. But once the migratory herds depart for the short grass areas of Loita and the Serengeti, the Mara lions once more turn their attentions to the resident prey animals. It is only then that the importance of land tenure to the various prides becomes clear. Only then do the lions have to work hard to feed themselves, killing buffalo, topi and warthog.

Once the rains begin in earnest, Musiara Marsh becomes a swamp. Then, only waterbuck and reedbuck, a handful of old bull buffaloes and the elephants dare to venture into its interior and the Marsh pride are once more forced to hunt the drier country above the spring line.

OVERLEAF *The reed-beds grow belly-high to a feeding elephant . . .*

1987 was a historic year for the Masai Mara. Senior warden Simon ole Makallah took the bold decision to close some of the most heavily used areas of the reserve for the duration of the long rains. No one was allowed to drive in the vicinity of Musiara Marsh, along Rhino Ridge or in an area near Keekorok. The ban was still in force when the migration arrived.

For many years people have expressed concern about the number of vehicles operating in highly congested regions. The ugly tracks they initiate are all too apparent, particularly during the rainy season when driving conditions can be appalling. A game-viewing plan, based on the construction and maintenance of a comprehensive network of all-weather roads, has already been outlined. Perhaps this is the only realistic solution to managing tourism in the reserve if even more facilities are built in the future, or existing ones expanded. In the meantime, the warden's decision can only be applauded by everyone who has the welfare of the Mara at heart.

With vehicles prohibited from entering the marsh and Rhino Ridge, drivers from Kichwa Tembo, Governor's, Mara Buffalo and Mara River tented camps concentrated their game viewing north of the reserve boundary, an area much favoured by the migratory wildebeest and zebra during the dry season. Here drivers could find all the animals visitors wished to see – except for rhino, which in recent years have been eliminated from this area by poachers. This is the home of a pack of wild dogs that range between the Mara River and Aitong Hill, twenty kilometres away. Like the cheetahs, the Aitong pack survives best outside the reserve, where hyaenas and lions – their major competitors – are fewer. And, should you want to see a leopard (as everyone does), there is no better place than among the thickets surrounding Leopard Gorge and Fig Tree Ridge. People soon realised that these northern rangelands, owned by Masai herdsmen – but freely shared with wild animals – constituted one of the best game-viewing areas to be found anywhere in the Mara.

. . . the home of a pack of wild dogs . . .

Though visitors may be stunned by the sight of the migration, it is not the strangely fashioned wildebeest they most want to see, it is the 'big five' – particularly the big cats. Perhaps as much as eighty per cent of game-viewing time is spent searching for or watching predators. They are the highlight of any safari, and the Mara provides them with unfailing regularity. But this quest to try and see everything in the space of just a few days takes its toll on the land. Areas favoured by the predators are constantly revisited. Over the years a maze of vehicle tracks has bitten deep into thickets, even penetrating the densest woodlands once a leopard's hide-out has been discovered.

This is well-illustrated by what has happened to the Gorge pride. These lions have prospered in recent years and have taken to frequenting a dense island of croton bush to the north of Leopard Gorge. It is among cover such as this that lionesses often choose to rear small cubs. Where before they were shy and hid from vehicles, now they make little effort to conceal themselves. The increase in

their numbers is perhaps a result of the abundance of resident prey animals, which now gather year round to feed among the northern rangelands. I often found the lions resting inside the thicket during 1986; it was a secure hideout for the lionesses and their numerous cubs, who spent endless hours scrambling among the cactus-like branches of a fallen euphorbia tree. By early 1987 vehicles had created a maze of tunnels through every part of the croton thicket, exposing it to a surge of encroaching grasses. Already you can see great patches of sky. Should the grass grow long through the wet season – as it did this year – and then burn during the dry season, it will eventually destroy the thicket and an important part of the lions' territory will have been lost.

The Isuria Escarpment rises high above the Mara plains, forming a natural boundary along the north west of the reserve. Looking east from atop the rugged scarp, I could understand why the Masai named this place Mara – the spotted land – for that is exactly how it looks from two hundred metres above the plains: in places stippled with a dense coating of thornbush.

Outside the reserve, young trees are abundant, woodlands and thickets flourishing. It is here that I search for leopard. But the regeneration of bush and forest in these areas does not please the Masai pastoralists, although they are part of the reason for it.

In the early 1970s, there were relatively few Masai settlements scattered around the northern boundary of the reserve. Now, there are numerous permanent villages. In the old days, the only habitations were the Masai's characteristic loaf-shaped houses endowed with flat, dung-covered roofs, huddled behind a protective circle of thornscrub, so well matched with the colours of earth and bush that one hardly noticed them. Today, you sometimes see spacious huts of mud and wattle protruding from the bomas, with thatched conical roofs in the style of the Kipsigis tribe. Scattered elsewhere are conventional-shaped buildings with bright tin roofs.

In earlier times, the Masai moved seasonally, responding in the fashion of true nomads by trekking with their herds between dry and wet season ranges, pursuing the rains – just as the migratory animals do. They had few permanent buildings. There was no need for them. I used regularly to see Masai families on the move, their meagre collection of possessions strapped across the backs of donkeys, their rangy cattle walking ahead of them across the open plains. Once the rainy season set in they moved on, allowing areas which had been heavily grazed to recover. Their impermanence was vital to the well-being of their land.

As they trekked from one area to another, the Masai regularly burned the dry grass to stem the tide of returning bush, preparing the ground for showers that quickly created a flush of green growth for their livestock. But the ever-increasing number of cattle grazing these areas now ensures that the grasses of the northern rangelands remain relatively short all year round. Ironically, this situation has helped to create ideal feeding conditions for the wild animals with whom the Masai must share their land. Many of the resident grazers – topi and gazelles, impala and kongoni – prefer to occupy the open plains for as long as they remain green, rather than venture into the sea of tall grass deeper inside the reserve. Added to these animals are the tens of thousands of wildebeest and zebra which annually migrate into this area from the Loita plains to the east and the Serengeti in the south.

The combined effect of domestic and wild animals outside the reserve has been to lessen the fuel available to burn back the encroaching bush. Driving through the regenerating acacia thickets in search of leopard already involves warring with a car full of tsetse flies. Perhaps eventually the noisy, biting flies, carriers of sleeping sickness, will return in sufficient numbers to reclaim the spotted land.

The Isuria Escarpment rises high above the Mara plains . . .

The Masai moved seasonally, responding in the fashion of true nomads . . .

During the drought of 1984, the wildebeest moved well beyond the Isuria Escarpment, reaching as far north as Kilgoris, further than ever before. The sight of all that free meat proved irresistible to some people, and poachers set their dogs amongst the wildebeest, then speared them, or trapped them with wire snares. Before long the local butchers were reporting a sharp decline in the demand for beef.

Usually by the end of October the last of the Loita wildebeest have departed from the Mara. But in recent years portions of their former range have been leased by the Masai landowners to farmers, and ploughed up for cultivation. Tens of thousands of hectares of wheat now cover much of the Loita plains, compressing the wildebeest and cattle on to areas such as these. In 1986 some of the migratory animals stayed well into the new year – some calved here in January and February 1987. When I departed in November, the wildebeest were still moving among the short green grass in their thousands.

Tolerance for wild animals has always been a feature of the Masai way of life – they have neither hunted game for food nor killed it for profit. Consequently some of East Africa's finest wildlife areas are to be found within the former range of these nomadic tribesmen. These attitudes evolved during a time when the wild herds did not seriously compete with the interests of the Masai pastoralists: the Masai moved into the Mara area in the mid-1600s. Now things are changing: populations are expanding, and the number of cattle has increased – people are hungry for land. It is no longer just for sport or to feed his dogs that a Masai occasionally spears a wildebeest calf; or that you sometimes see a group of young warriors chasing a herd of wildebeest from their grazing lands. The Masai naturally resent the burden of all these wild animals and despair as they see the grass eaten and trampled within a few days of the wildebeest's arrival. And they question why their cattle should be forbidden from sharing the grass and water denied to them within the reserve boundary. Echoing these sentiments, perhaps, in 1984 the government acknowledged the demands of the Masai and degazetted 162 square kilometres of the Mara from protected status, reducing the size of the reserve by ten per cent.

While poaching in the Mara has declined sharply during the last few years, there has been a dramatic increase in activity in the Serengeti. Many elephants previously living in the northern woodlands now seek refuge across the border in Kenya.

Elephants are highly intelligent creatures. They soon learn to seek shelter in protected areas, avoiding conflict with man where possible. Consequently the impact of the elephant herds is far more evident within the parks and reserves than outside their boundaries. But as suitable food sources become exhausted, elephants are forced to wander further and further from the safety of protected areas.

Apart from the thousand-plus elephants living within the Mara Reserve, there are already an additional four hundred moonlighting in the surrounding areas

The Masai have neither hunted for food nor killed for profit . . .

such as the northern plains. It is not uncommon to encounter herds of up to a hundred elephants tearing up the acacia thickets around Leopard Gorge and Fig Tree Ridge, or trundling back down well-worn trails along the Isuria Escarpment after plundering some favoured food resource during the night. For the time being these elephants survive in a benign environment, generously tolerated by the Masai herdsmen whose land they share. It is a wonderful sight to see a family of elephants at dusk boldly outlined against the distant horizon as they wander peacefully from one patch of thicket to another, silently making their way across the open plains not far from Aitong. Yet such vehicle-habituated giants may one day prove easy targets for the automatic weapons of motorised poaching gangs. In northern Serengeti, they already have.

7 *Surviving the River*

To see ten thousand animals untamed and not branded with the symbols of human commerce is like scaling an unconquered mountain for the first time, or like finding a forest without roads or footpaths or the blemish of an axe.

BERYL MARKHAM – WEST WITH THE NIGHT

Born in a shallow swamp among the dwindling forests of the Mau Escarpment, the Mara River flows like a great artery for fifty tortuous kilometres through the reserve. As it courses from north to south, it draws to it a network of tributaries that provide every corner of the reserve with water. Talek and Ntiakitiak, Olare Orok and Olkeju Ronkai: Masai names for the water-courses which in earlier years provided seasonal water for the tribesmen's thirsty cattle. Officially the Masai are now prohibited from bringing their cattle into the reserve, but in a dry year hardly a day passes without cattle grazing and watering inside the boundary.

The river is like a chameleon: in places it flows calm and unhurried between steep muddy banks; elsewhere it rushes over great chunks of black rock, a deafening white-water torrent avoided by its larger residents, hippos and crocodiles, which prefer calmer resting places.

In a normal year, the river is an unfordable barrier separating the eastern half of the reserve from the Mara Triangle in the west. Occasionally, in times of drought, it is possible for a Toyota Landcruiser to edge carefully down the steep banks of the river and brave the maze of half-submerged rocks which provide an uneven footing on which to cross. The only permanent crossing points are two narrow concrete bridges: one just to the north of the reserve on the road between Mara River Camp and Kichwa Tembo; the other a few kilometres west of Keekorok Lodge on the road from Serena.

During the long rains the river sweeps all before it. Trees that have lain secured on the muddy bottom are carried downstream on the flood-waters, stacking up in massive piles against the low-slung northern bridge. If the torrent rises further, vehicles are forced to wait for hours before they can pass in safety. As night draws in, Masai herdsmen may be tempted to take their chances, clinging to the tails of their scrawny cattle as they struggle to avoid being swept away in the white water.

Having run its course through the reserve, the river cuts sharply west through the northern Serengeti and eventually empties into Lake Victoria. In this manner it partitions a triangle of rich dry season grazing, bounded to the south and east by the river, and to north by the Isuria Escarpment. To enter the Lamai Wedge and Mara Triangle, the wildebeest and zebra must brave the river.

Coming over a rise just east of the Mara Bridge on the road to Keekorok, there is a clear view of the Serengeti barely a kilometre away to the south. It feels almost as if you could reach out and touch it. During the years when the border was

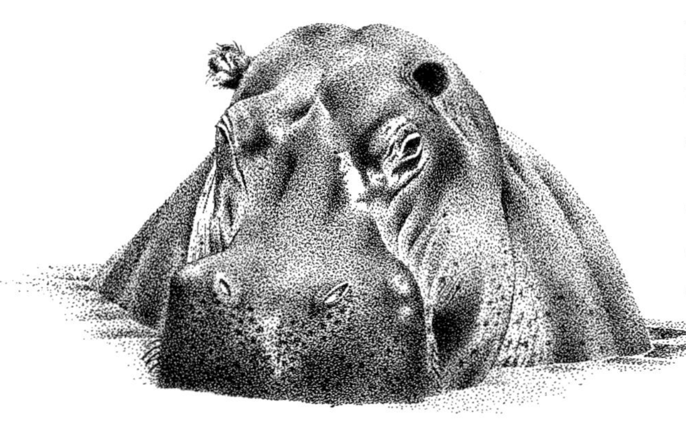

The river flows calm and unhurried between steep muddy banks . . .

closed, this was as near as I could get to the Serengeti. I would sometimes drive to that spot and look down into the beckoning land and long to be there: to see for myself where the wildebeest were massed. And occasionally during July or early August, I would be lucky enough to find thousands of wildebeest streaming from the hillsides of the northern Serengeti to join those already crossing a wide stretch of the Mara River. It seemed so close, yet so far, denied to me by the invisible boundary separating the Mara from the Serengeti.

I have never discovered why certain crossing places become imprinted in the minds of the wildebeest. Sometimes they try and cross the Mara River at suicidal places, plunging over sheer cliffs and drowning in their hundreds beneath an

Horned heads rear out of the water, eyes bulging with panic . . .

The river narrows with a gruesome litter of dead bodies . . .

OVERLEAF *Dark thunder-heads ballooned across the sky . . .*

The wildebeest streamed south towards Musiara Marsh . . .

insurmountable wall of mud. But what appears to be a difficult crossing place this year may have been far easier in the recent past. The geography of the river banks is constantly changing. There may be a historical perspective underlying the wildebeest's choice which is not apparent when viewed on a smaller time scale. Older, more experienced animals probably return to places where they have successfully crossed before. Regardless of how many other wildebeest died there in previous years, *they* survived to cross again.

Certain crossing places attract the largest number of animals because the nature of the terrain makes them clearly visible from a long way away and creates the setting for a massive exodus of animals across the river. Though the wildebeest and zebra on the surrounding plains and hillsides may not be able to see the river itself, the lie of the land enables them to see other animals heading in that direction. Hippos are continually eroding new exit points, making trails used by other animals that come to drink. Each animal in its own way helps to modify the contours of the bank, as does the river itself. Use begets use: over the years the vegetation is stripped and trampled at these places, helping to create an even more visible – and safer – crossing site for the herds.

Whatever the factors may be that prompt the wildebeest to cross – and sometimes it is simply a consequence of a build-up of animals wanting to drink – nothing deters them once the urge is established. If vehicles or predators disrupt a large crossing at a favourite fording place, the wildebeest simply move on to the next, or establish a new one, even thundering through thick forest to reach the river if need be.

There is one particular stretch of river separating Paradise plain from the Mara Triangle to which the wildebeest and zebra return every year. On a hilltop to the west you can clearly see the manyatta-styled rooms of Serena Lodge perched high on a ridge above the river.

The first to use these crossing sites are the migratory zebra. They chomp their way through the tall red oat grass surrounding Paradise plain earlier in the year than the wildebeest. Their river crossings are much more cautious and orderly. You rarely see the carcass of a drowned zebra. The greatest danger for a zebra is not of drowning but of falling prey to a lion or a crocodile. One year a large male leopard staked out this area and ambushed the zebra families. Over a period of two weeks he picked off a number of small foals as they emerged from the water.

The two most commonly used crossing places in this stretch of the river are only a few hundred metres apart. One is relatively open, providing good visibility where the river runs rapidly in a spill of white water over black rocks. But the other is quite different: a cul-de-sac of open ground, hemmed in between the river and a dense thicket of cover, where the Paradise lions often lie in ambush.

The banks are steep on both sides, except for one stretch of mud flats where the river bends sharply. It is here, on the flat muddy terrace, that the wildebeest gather to drink or cross, and when they do, half a dozen crocodiles will already have entered the water in anticipation.

. . . moving in single file, silhouetted against the rising sun . . .

But despite the lions and the crocodiles and the five metre wall of mud facing them, the wildebeest remain undeterred. Each year is the same: a crush of animals desperately treading water on the far side of the bank, searching for a way out. Hundreds of horned heads rear out of the river, eyes bulging white with panic. Usually the first animals are swept back to the shallows unharmed. But by then others have galloped into the water behind them, forced forward by the swelling tide of wildebeest trying to enter the cul-de-sac. As more and more seek to cross, the dust rises into the sky like a smoke signal, encouraging still other wildebeest to hurry from the plains and follow. The animals seem

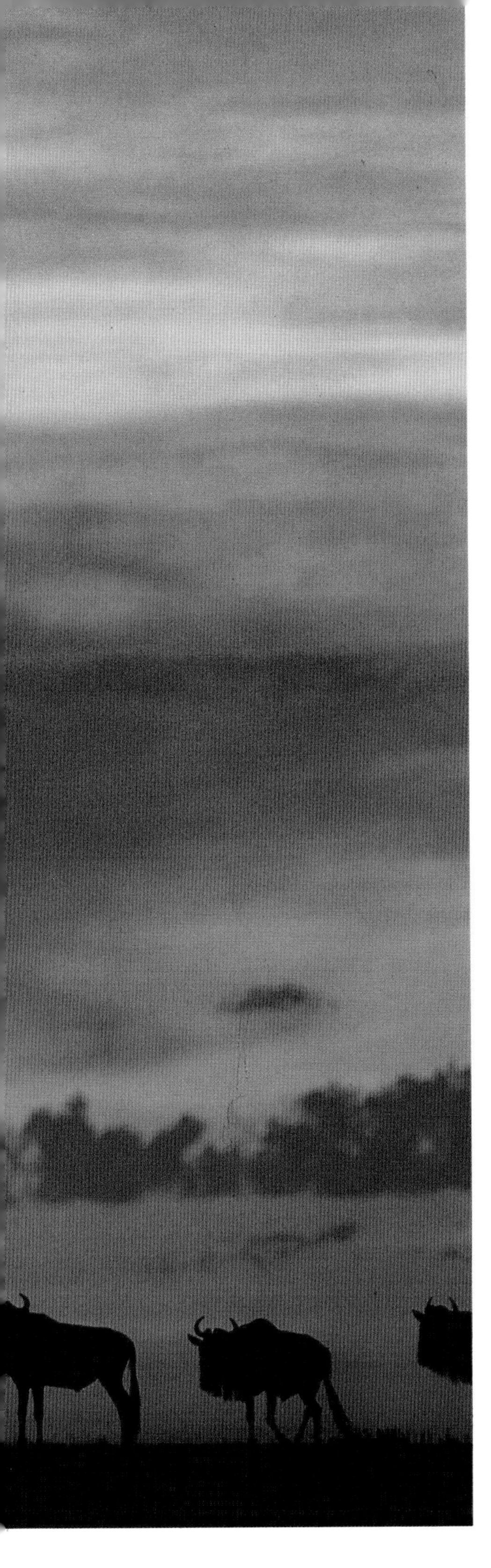

overtaken by the moment; soon the river turns black with a confusion of thousands of grunting, bellowing wildebeest, and the water heaves with bodies.

If the build-up of animals trying to cross is too great, scores of wildebeest are trampled or drowned. Others are swept further downstream by the current and become tangled among the branches of fallen trees. But despite these losses, the majority eventually find a way through. Somehow they manage to pick their way along the steep rim of mud, until a hippo trail or another natural break in the bank releases them.

Many calves become separated from their mothers during the pandemonium of the more difficult crossings. As the bulk of the herd heads off into the distance, scores of cows and calves gallop to and fro, grunting and bleating, mingling with those still trying to cross, sometimes even re-entering the river in their efforts to relocate each other.

Some years are worse than others: the height of the river, the place they choose and the number of animals crossing at any one time will determine how many animals die. In a bad year, the Mara claims thousands of victims, and the course of the river is made visible from kilometres away by the constant swirl of vultures overhead.

Each day in mid-September, at around the time when visitors were arousing themselves from their afternoon siesta, dark thunder-heads ballooned across a sky turned purple by the dying sun. By four o'clock the first heavy spots of rain were pelting down.

Initially when the showers continued day after day, people talked of 'grass rains' – a spate of scattered showers sometimes experienced before the short rains begin in earnest in mid-October. Call them what you will, it soon became apparent that rains had arrived. While the rest of Kenya remained dry and dusty, the Mara Triangle soaked up the water like a sponge and converted it into a broad swathe of short green grass.

On the other side of the Mara, conditions remained inhospitably dry. In early September a huge fire swept into the reserve from the east. In the space of just a few days it removed the last of the dry season forage around Keekorok Lodge, severely scorching the forests and acacia thickets. Fuelled by grass grown long during a prolonged wet season and burning at the height of the dry season, these are the most damaging of the fires. Thousands of hectares of life-sustaining forage were lost in this manner, and when no rain fell, barely a wildebeest or zebra was to be seen in the eastern part of the Mara.

In Serengeti, the scientists and parks management have joined forces to implement a fire policy to try and prevent just such occurrences, and to help promote the regrowth of the dwindling northern woodlands. Once the rains are over, they initiate a series of early burns while the grass is still relatively green. These burnt areas help to prevent the unimpeded spread of fires started later in the dry season by poachers and cattlemen. But even with the best intentions,

OVERLEAF *Nothing deters them once the urge to cross is established . . .*

LEFT *The first to use these crossing sites are the migratory zebra . . .*
BELOW LEFT *The greatest danger for a zebra . . .*

things may not work out according to plan. Vast areas of the northern Serengeti were burned this year as extensive fires ravished the investment of the long rains.

Encouraged by the rainy conditions elsewhere, the wildebeest streamed off the northern rangelands and headed south towards Musiara Marsh. They did not abandon the area in a single dramatic exodus in the span of only a few days, but departed in a steady stream over the course of weeks. Early each morning I would meet them moving in single file, silhouetted against the rising sun.

Having fed and watered around Musiara Marsh, the herds pushed further south, ignoring the crossing places that earlier in the dry season had facilitated their passage from the Kichwa Tembo area to the northern plains. Though they congregated at these same places to drink, they rarely used them to recross the Mara River on their return journey to Serengeti – the Kichwa Tembo plains no longer looked green and inviting. Instead they headed for Rhino Ridge and Paradise plain, reducing the rank stands of dry grass to stubble as they passed through. Others moved south east towards the Talek area, avoiding the Mara River altogether.

. . . is of falling prey to a crocodile . . .

My heart quickens when a lion emerges from cover . . .

Some people expressed disappointment, their perception of the migration coloured by memories of drier years, when the animals arrived and departed in greater concentrations. But the drivers from the tented camps for once welcomed the wetter conditions and greater availability of grass. In previous years they had had to placate visitors who had arrived too early or too late to witness the river crossings. This year was different. The wildebeest were so scattered throughout their northern range that when they departed they crossed day after day, week after week, slowly seeping back into the Mara Triangle. But not without losses.

The great drama of the river crossings is heightened by the presence – real or imagined – of the predators. At any moment a lion or leopard might explode from its hiding place towards the waiting herds. In the river itself, unseen except for the frantic splashing of its prey, a crocodile drags its victim below the surface.

It is not for reasons of glorifying violence, or blood thirstiness, that my heart quickens when a lion emerges from cover. Nor is it a voyeurish desire to witness sudden death or hear an animal cry out. It is a fascination with the hunt, an unconscious stirring of some inner self: an atavistic echo from the time when man the hunter first roamed this land. It is the chance to watch two perfectly fashioned creatures: one adapted to flee or hide, the other to pursue and kill.

Some days I scarcely bother to eat when searching for something new to photograph, so anxious am I not to miss the best of the light. It never seems quite the right moment to stop for breakfast or dig into a picnic lunch. There is always some impending drama just over the next rise. Perhaps lions battling to overpower an old bull buffalo, or a leopard nursing her cubs.

Over the years, I had managed to photograph all the larger predators – except for crocodiles. They remained exceptionally shy and for as long as there were few vehicles roaming the edges of the river, it proved impossible to get close to these giant reptiles. All that I might find would be a tell-tale ripple, an arrow-shaped bow wave pointing from the river's edge to the barely visible snout gliding effortlessly through the water.

. . . a leopard nursing her cubs . . .

wned wildebeest floated on downstream . . .

But slowly, as the number of vehicles increases, the crocodiles have become more tolerant. Mouths agape, they now sun themselves on the mud banks at the edge of the shallow waters, or bask mid-stream on low rocky ledges, unmoved by the excited exclamations of the visitors.

Crocodiles conjure up visions of our prehistoric past, having survived virtually unchanged for 190 million years as the only reptilian descendant of the Archosaurs – a group that included the dinosaurs. There is a strange beauty in the colour and texture of a crocodile's wet skin, its coat of armour gleaming green and yellow as it emerges from the shallows. We fear them for their brute strength and unreasoning violence; yet feel compelled to seek them out.

Growing up to six metres in length and weighing three-quarters of a tonne, the largest of crocodiles is the most formidable of predators. As the lions and hyaenas dominate the predatory hierarchy of the savannahs, so the crocodiles reign supreme in the tropical fresh waters. One man-eating crocodile that had been shot and gutted revealed the remains of a full-grown leopard in its stomach. Even an animal the size of a Cape buffalo may suddenly find itself hopelessly out of its depth when seized by a large crocodile. Except for the elephant and adult hippo, nothing is too big or too fierce to escape the terrible jaws of the great saurians.

I drove carefully along the river bank. The tension that had built up inside me eased. For once, there were no other cars. I was alone to savour the wilderness, free to merge into the scene and become a part of it. Below the banks, the dark brown shapes of drowned wildebeest bobbed through the rapids, then floated on downstream. There were dozens of them.

Day after day, the wildebeest had been massing at their favourite crossing places at the southern end of Paradise plain, trekking from morning until evening across Rhino Ridge, leaving the northern acacia thickets and Musiara Marsh far behind them. The grass had long since vanished. Where the herds stood waiting, the ground was churned to dust. Thousands crossed each day, and each morning I arrived to watch them depart. From six kilometres away on top of Rhino Ridge, I could see the tight columns emerging silver-backed from the water and moving resolutely away into the Mara Triangle. Wave after wave of wildebeest streamed south towards the Serengeti.

I moved on until a dark stirring on the far bank caught my attention. The light had begun to fade and at first I failed to understand what was happening. Gradually, the dark shapes transformed from rock to wildebeest, and from half-submerged log to enormous crocodile: the same giant reptile that I had often watched lying motionless along this familiar stretch of river. Like a monstrous gin-trap, its jaws gaped open, then smashed shut: 66 conical teeth reaching out to grab the shoulder of the fallen wildebeest.

Slowly, the crocodile eased back into the water, digging into the muddy bottom with its clawed feet and paddling backwards with its tail. The wildebeest bull –

Its jaws gaped open, then smashed shut . . .

weighing more than 180 kilogrammes – struggled to regain its feet. If only it could gain just a few more metres, it might reach dry ground and be able to break free. But the task of hauling something more than treble its own weight was beyond it.

The bull fell to its knees at the water's edge. For a moment the crocodile let go and lay half-submerged alongside it, sensing that its victim was already too weak to escape. A minute later the crocodile glided forward, heaving its upper body out of the water. Twisting its jaws sideways it drove its pointed teeth into the wildebeest, tearing a great hole in its side. I watched, stunned by the violence of the crocodile's actions, forcing myself to accept something which was so natural, yet so appalling to the human way of thinking. Again and again the wildebeest struggled to free itself from the nightmarish scene. And each time it did so, the crocodile launched itself through the water like a battering ram, torpedoing into the wildebeest and pulling it down again. At one point the bull almost gained its feet, scrambling over the rocks on its knees as the crocodile held it back by the tiniest piece of skin. I willed it to break free, yet knew that its injuries were already beyond healing.

Attracted by the blood of the dying animal, dozens of catfish seethed around the crocodile's armoured head, competing for scraps of flesh. Normally these predatory fish contribute to the diet of large crocodiles, but this one could afford to ignore them while there was easier prey to eat. Suddenly the crocodile seemed

. . . creating abstract images like ancient rock paintings . . .

to notice the tide of bodies floating past. It turned and swam to mid-stream, intercepting one of the drowned wildebeest and guiding it towards the bank. Within the last hour more than three hundred wildebeest had drowned in the cul-de-sac.

Some of the survivors of this latest crossing turned west, and in a single column trudged wearily along the top of the bank, moving towards the spot where the dying wildebeest lay. The herd paused briefly above the bull, staring down at the stricken animal. In a pathetic enactment of the most basic instinct in its life, the bull gathered itself and struggled to its feet, as unsteady on its spindly legs as a new-born calf. It careered about wildly, desperately trying to clamber up the steep bank – to follow, to rejoin the herd. But it was hopeless.

Next morning I returned. Two hyaenas feasted on the half-eaten carcass, while restless vultures crowded around, impatient to finish the job. By afternoon nothing remained to distinguish the wildebeest's clean white bones from those of other victims of the river crossings.

A deathly hush stilled the air, as if the land were in mourning. Marabous and a solitary goliath heron stood solemnly, silhouetted against the overcast sky. Crocodiles rested like scaly boulders among the dead, basking on slabs of coal-black rock. All activity ceased. Even the vultures fell silent,

It was like some awful re-enactment of a scene from the World War I trenches: scores of bloated bodies slumped awkwardly against one another, half-buried in the khaki-coloured mud. Some of the wildebeest lay entombed in the morass, creating abstract images like ancient rock paintings. Occasionally a trapped animal would stir, clinging stubbornly to life, thrashing its legs wildly in a feeble attempt to free itself from the deadly combination of mud, animals and tree roots.

The vultures watched impassively with dark beady eyes. The stench was overpowering. A pied wagtail looked almost unnaturally clean as it bobbed and weaved, deftly feasting on the hordes of flies that clustered over the carcasses to feed and lay their eggs.

Visitors that chanced upon the sombre scene expressed horror at the suffering endured by the trapped creatures. Some demanded that ropes and chains be mustered to rescue the dying survivors, unwilling to accept the event as an inevitable part of the natural order of life on the plains. Human emotions seem strangely out of place in such circumstances. Deaths by misadventure like these are in stark contrast to the senseless killings precipitated by man's deliberate acts of aggression.

Many of the imprisoned victims of the river crossings sustain serious injuries – broken legs, damaged spines – and would be unfit to continue their journey after being pulled from a graveyard full of bodies. Some people have proposed that difficult stretches of river bank be hacked open, or that ramps be built to make easier crossings for the wildebeest, thereby compensating in some measure for

Vultures probed and prodded . . .

the tens of thousands of animals killed by poachers each year.

But it is not events such as these that ultimately control the size of the population. The migratory wildebeest have been crossing rivers for thousands of years, have died all manner of deaths, then replenished their numbers with the birth of hundreds of thousands of calves. If man tampers with the system, nature will simply redress the balance elsewhere. If more animals survive the river crossings, more will succumb to the ultimate controlling factor – the availability of dry season forage. Do we then start shooting wildebeest that would otherwise die due to the effects of under-nutrition? Far better to leave them alone rather than blunder around in the wilderness of our own ignorance.

The river soon washed itself clean again. Crocodiles and hyaenas, monitor lizards and catfish all played their part in completing the cycle. Vultures probed and prodded, ripping out the softest tissues, and within a few days the bloated carcasses collapsed, leaving nothing but mud-spattered skin draped around the bony wrecks.

Having finished feeding, the vultures soared high above the cul-de-sac, where they could see the last of the wildebeest huddled together at the cobbled entrance to the crossing point. Gone was the boldness of the larger herds. A perilous journey lay ahead of these last few animals, unshielded by the massed departure of earlier crossings. Those who had crossed unscathed moved purposefully away from the river and spread out across the plains.

As the afternoon storms continued to soak into the dark soils, a wave of shoots sprouted anew where the wildebeest had grazed earlier in the year. Likewise, the land between Isuria Escarpment and the main road to Serena Lodge – burnt earlier in the dry season by the Masai – beckoned once more to the wandering herds. Areas of long grass could now be ignored or abandoned, though for a while the zebra continued to bury their faces deep in the knee-high vegetation.

Before long all manner of animals gathered to feed on the rich short grass areas: warthogs and baboons, Coke's hartebeests and topis, impalas and Thomson's gazelles, as well as thousands of wildebeest and zebra. Winged termites were released from their earth castles as the showers came and went, providing a new generation of insect kings and queens, though the majority would be consumed by jackals and bat-eared foxes, eagles and hornbills, mongooses and monkeys.

In effect, the wildebeest had prepared the land for their own departure. Many of the animals that I now watched had undoubtedly passed through these areas once already on their journey north from the Serengeti, in late July and early August. At that time, I had seen tens of thousands of them successfully cross the Mara River near Kichwa Tembo, as they hurried onwards through the acacia thickets north of the marsh to reach the plains beyond – only handfuls had drowned at these easier crossing places.

Over the next few weeks more and more animals re-crossed the Mara River to feed in the triangle, crowding into the area until their tracks cut shallow paths over every part of the grasslands. For as long as the rains continued, the wildebeest stayed, mirroring the extent of the green flush with tens of thousands of dark bodies. It was too early yet for the animals to venture far to the south. The Serengeti's plains were still dry and grassless, without any permanent source of drinking water for the thirsty herds. The wildebeest would bide their time until the rains of October were well underway before hurrying back to their wet season pastures.

But why leave at all when the rain is falling and the grass green and growing? The Mara appears to have everything that a wildebeest might need during wet or dry season. Why risk yet another river crossing, then run the gauntlet of poachers and predators so as to be back on the Serengeti's short grass plains before the beginning of a new year?

Scientists at the Serengeti Wildlife Research Centre have long been working to solve one of the most intriguing mysteries of the wildebeest's annual journey.

Impalas gathered to feed on the rich short grass areas . . .

For at least a million years the migratory herds have probably spent the wet season on the Serengeti's short grass plains. Thirty years ago, when the herds were much smaller than they are today, wildebeest only needed to move as far as the Western Corridor to survive the dry season – they did not travel north as far as the Mara Reserve. And they always returned to the plains once the rainy season set in.

Certainly the open plains are a safer environment for the herds than the woodlands and the long grass areas. On the plains they can see what is happening around them, and avoid the predators more easily. During the long rains the Mara's heavy soils become waterlogged, creating conditions underfoot that the wildebeest tend to avoid. It is the character of the Serengeti's volcanic soils, and the grasses that flourish on them, which beckon to the wildebeest, drawing them back each year. Researchers have discovered that the red oat grass which dominates so much of the woodlands in the north and west is not adapted to withstand heavy grazing and has only moderate levels of calcium – a vital nutrient needed in relatively high concentrations if a cow wildebeest is to produce sufficient milk for her calf without loss of condition. The short grasses are rich in calcium – as is the water on the plains – and better able to sustain intensive grazing. The grasses have evolved side by side with the wildebeest – theirs is a partnership in which both can flourish. But if the migration were to stay too long in the long grass areas – or return too often – they might deteriorate. And so, each year, the herds move on from the Mara, and return to the Serengeti.

 . . . mirroring the extent of the green flush with tens of thousands of dark bodies . . .

I drove steadily for an hour along the base of the Isuria Escarpment. Wherever I looked I saw wildebeest. Wildebeest feeding on the open expanses of short green grass; wildebeest standing or lying beneath the shade of the taller acacia bushes; and still more wildebeest picking their way nimbly down the rocky pathways of the rugged escarpment. Some of the animals were grazing with their heads facing towards the north east, but the majority moved slowly south in the direction of Serengeti.

Wastepaper plants littered the grasslands like white confetti tossed casually across the plains. Warthogs accompanied by clusters of tiny piglets knelt to crop the welcome green shoots. Nearby, a small group of topi females posed elegantly with buff-coloured calves among thick stands of thornbush too tall now to be destroyed easily by dry season fires. The arrival of young animals such as these was keyed to the onset of the rains. The time had come for the migratory herds to follow the rains south again.

High above me a cascade of white water raced headlong down the steep rock face of the escarpment, emerging again as a brown torrent to nourish the plains and provide water for the passing herds. The sound of distant rushing water drowned the forlorn bleating of a lone wildebeest calf. Though old enough to feed by itself, it looked pathetically vulnerable away from the shielding presence of the herd. I could not help wondering if it would survive its first migration year, somehow outwitting the predators that even now were perhaps watching its every move.

Bird song mingled with the rising humidity until at last the storm clouds burst open, drowning my thoughts as the rain hammered on the roof of the car. The wildebeest feeding on the higher ground turned and hurried off the slopes, then stood facing south, like some vast legion of soldiers called to attention on a huge parade ground, their backs hunched against the elements, glistening wet and grey.

Ahead of me, high above the trees, I could see the water tower and white flagpole of the anti-poaching camp at Ng'iro Are, 'the place of brown water' – on the border between Kenya and Tanzania. In the distance lay the blue hills of Serengeti.

Hundreds of thousands of wildebeest had already passed this way. One last river crossing lay ahead of them, where the Mara flows wide and shallow through the Serengeti before finally emptying into Lake Victoria. If the rains faltered the herds would wait at the edge of the woodlands; if they continued, the wildebeest would reach the plains by December.

The sun broke through the thick cloud cover, splashing shafts of light on to the green surrounds. I stopped for a moment, then turned back towards camp. My migration year was over once again. The wildebeest were headed home to their ancestral calving grounds.

 . . . free to roam across that huge, dusty continent . . .

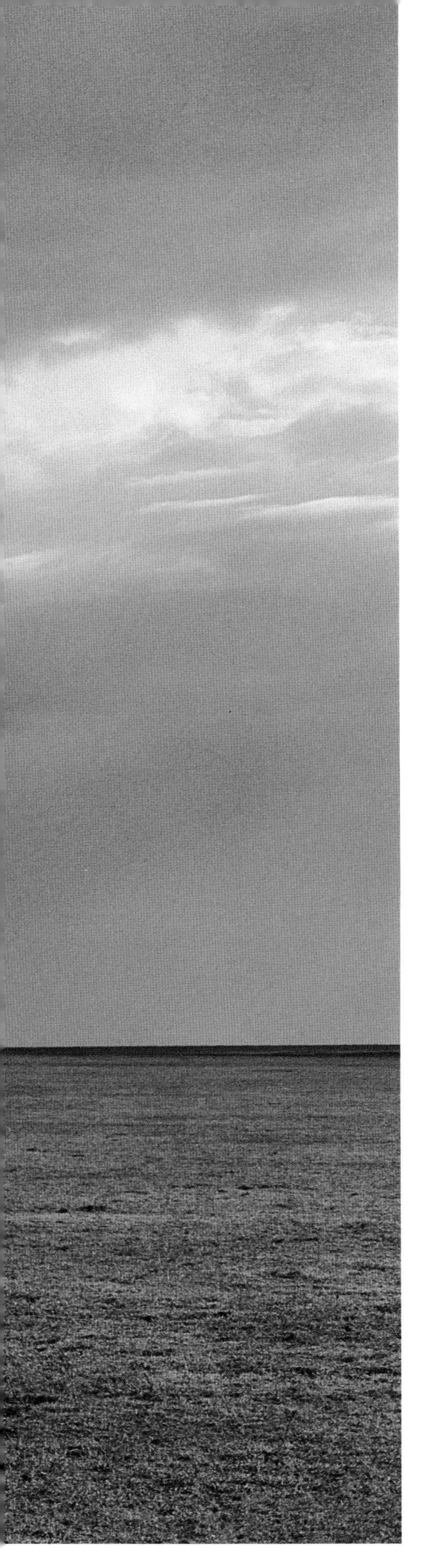

Epilogue

The ritual never changed; the buffalo always came north. It was described as the grandest spectacle in all of nature, the juggernaut advance of millions of great creatures across one quarter of a continental wilderness. There the animal was triumphant over its environment, so successful that a beast became the architect of the grasslands, with weather and soil its servants.

FRANKLIN RUSSELL – THE HUNTING ANIMAL

North America's great herds have gone, sacrificed to man's insatiable greed. Today only a few thousand animals remain as a reminder of the passing of the greatest wildlife pageant the world has ever seen, a time when perhaps thirty million bison roamed the prairies.

Spectacular herds once gathered in many parts of the world, as witnessed for instance in prehistoric French cave paintings. But Europe and America, India and the rest of Asia, now stand silent by comparison to their past animal wealth. It is incomprehensible that so much life could have vanished so quickly.

A hundred years ago in Africa, the animals were still free to roam across that huge, dusty continent, unrestricted by extensive human habitation. Early settlers in southern Africa described how tens of thousands of springbok journeyed en masse across the land: 'Over a hundred miles long by fifteen miles wide was covered by the trekbokken moving in an unbroken mass giving the veld a whitish tint, as if covered with a light fall of snow.' At the other end of the continent, gazelles seasonally trekked through the arid wastelands of the Sahara. Hartebeest moved in their thousands through parts of East Africa, and ivory hunters told stories of huge bands of elephants that migrated annually across the mountain country of northern Kenya. But these are visions of our past, reflected in the twilight of the long African day. Only the wildebeest migration of the Serengeti and the 800,000 strong army of white-eared kob in the south-eastern Sudan now serve as a reminder of those days.

For the last 50,000 years, man has increasingly distanced himself from his animal origins, bringing about change through cultural and social evolution alone. Dramatic advances in technology have freed us to alter drastically what still remains of our natural environment. But by increasingly isolating ourselves in the bleak world of cities, we have lost touch with our biological base. The ultimate price of material comforts is the destruction of wilderness.

It is perhaps naive to suppose that mankind will cherish and stand guard over places like Serengeti and Mara for their own sake. For every Serengeti, there is another unique wilderness perishing through lack of support. Yet Africa's wild places are ancient monuments, preserving a priceless fragment of Pleistocene life, a memory of how it was before man assumed a position of dominance over his fellow creatures. Animals need space of their own. It is their right.

In his introduction to the British edition of *Serengeti Shall Not Die*, Alan Moorehead wrote that 'the Serengeti is supposed to be a reserve where no wild animals can be killed, but there is insufficient money to police it properly, and

OVERLEAF *Is it fair to expect African countries to set aside vast tracts of land?*

the demand for the land for agricultural purposes is steadily increasing everywhere.' Such words are as true today as they were thirty years ago. Man has shown quite clearly and ruthlessly that there is no guarantee of life for the other animals, inside or outside protected areas; no place safe enough for rhino or elephant so long as there is a price on their head. To survive, the parks and reserves must pay for themselves.

The Grzimeks helped publicise the importance of places such as the Serengeti and to enlist much needed support in Tanzania and overseas. In the wake of their efforts, and those of the government, and with the publicity the Serengeti received from books and television, tourists began to flock to Tanzania's parks and reserves. Bernhard Grzimek always believed that the revenue generated by tourism would help prove that maintaining wildlife areas could also be economically viable – though I am sure he would have been the first to add that the developing tourist industry should make it a priority to help safeguard the habitat and the well-being of the animals.

At the time of the border closure in 1977, Solomon ole Saibull, Tanzania's Minister for Tourism and Natural Resources, had this to say when reviewing conservation efforts in the early 60s: 'At this time one man did a tremendous amount to bring this idea of conservation through tourism to the notice of the world. This was Professor Bernhard Grzimek, whose many services in the early days – and now – deserve the highest possible honour. His book *Serengeti Shall Not Die* was less a reassurance than an expression of the determination of a desperate conservationist. This, and the subsequent film, became a symbol for conservation – not only relating to the Serengeti, but to all other areas of natural attraction.'

When I first decided to follow the story of the migration I was very aware of the dilemma I would face in comparing the Mara with the Serengeti. Visiting the Serengeti was like a breath of fresh air. But that was only to be expected. Having lived in the Mara for most of the past eleven years, I found Serengeti new and exciting – it was a journey of rediscovery, as if I had only just set foot in Africa.

It would have been all too easy to sing the praises of the Serengeti while pointing out the problems that the Mara struggles to resolve. Yet they are the same issues which the Serengeti is now having to deal with: how many camps and lodges to build; when and where to allow off-road driving; how to deal effectively with animal harassment. And there will always be the poaching problem.

The border closure in 1977 provided Tanzania with a temporary and unintended reprieve from the harsh realities of wildlife conservation. A reduction in the number of tourists helped poaching to flourish. Car-loads of visitors not only help to finance conservation, they create a presence in areas where poachers might otherwise operate unhindered. The government now has the chance to reap the benefits of tourism without allowing it to be overly harmful to the environment.

The Mara is no longer a little known sanctuary. Today it is recognised for what it always has been – Kenya's finest game-viewing area. In the knowledge of this, the authorities now seem determined to try and give the Mara the protection it deserves. Earlier this year the government announced a moratorium on the building of any new lodges or camps in the vicinity of the reserve. And in an effort to control vehicle pressure – particularly around predators – three Suzuki jeeps have been donated through the World Wildlife Fund by the Mara Trust. The jeeps will be deployed in popular game-viewing areas to try and control the impact of tourism on the animals.

This year the Aitong wild dogs established their den a few kilometres from Mara Buffalo Rocks, where drivers regularly search for leopard. Once the dogs had been located, the area became riddled with vehicle tracks, some of which passed within a few metres of the den. To protect the puppies the Director of Wildlife ordered the area temporarily closed to all vehicles. Measures such as these promise to help redress the balance between man and the animals.

So at present there is room for optimism. Negotiations are underway between the wildlife department and the Masai to try and safeguard the northern rangelands, a haven for resident and migratory animals which also harbours all the larger predators. The Masai landlords receive little direct revenue from the heavy density of tour vehicles now using this area. It has been proposed that some official status be given to the northern rangelands and other areas surrounding the reserve on condition that the land is not sold or leased for settlement or developed for agricultural purposes. This Conservation Zone would act as a buffer, safeguarding the grazing rights of the Masai and protecting the habitat for the abundant wildlife. If the Masai agree to the proposal they will receive a daily fee for each person visiting the area. Much better than having to watch car-loads of wealthy visitors criss-crossing every corner of their land, yet contributing little to their personal welfare. And some compensation perhaps for continuing to tolerate the presence of all those wild animals who compete with their cattle for grazing.

Americans and Europeans fly thousands of kilometres to mourn the loss of their own natural heritage by visiting what remains elsewhere: it is the outside world that so wants to visit Africa's wild places. Few Africans can afford the luxury of spending time watching animals, or enjoying the timelessness and beauty of their own land. Many have never seen a wild lion or an elephant. And those who have sometimes pay the price – the killing of their livestock or the destruction of their crops. Animals are not to be deified. Is it fair to expect African countries to set aside vast tracts of land for the benefit of wildlife, and to finance them as well? Surely the whole international community must help in ensuring the survival of places such as these.

Conservation is an expensive business. Wilderness areas can only be truly viable if they are self-supporting and able to hold their own against competing

Many Africans have never seen a wild lion . . .

forms of land use. Otherwise their future will never be secure. If conservation is to succeed in the long term, the local people must benefit in some tangible way. Their co-operation can only be guaranteed through incentive and involvement. Not through coercion.

When Bernhard and Michael Grzimek first saw the Serengeti, they were bewitched by the land and its vast assemblage of animals. They perceived the Serengeti as hallowed ground, a tract of primordial wilderness where the rest of creation was on equal footing with mankind. And they were right. There is nowhere else on earth where you can see such an incredible array of animal life – more than two million ungulates, together with the predators that feed on them. Here the game has survived unchanged for thousands of years, despite the periodic devastations wrought by disease and drought.

The Grzimeks were determined to help ensure that the Serengeti – and places like it – should be protected for all time. Bernhard Grzimek died in 1987, while I was still writing this book, having dedicated a lifetime to the pursuit of a dream: that Serengeti should not die.

I have travelled the length and breadth of North America, visiting places renowned for their natural beauty. And I have looked out across the prairies and been filled with a sense of great sadness: how could we have let so much animal life disappear into oblivion? But my journey through Serengeti/Mara filled me with hope. There *is* still so much here. We *do* still have time. Are we really prepared once again to bear silent witness to the loss of something as grand as the great migration? I have no doubt that we are, unless we remain constantly vigilant. The efforts of people like David Babu, Myles Turner and the Grzimeks urge us not to let it happen again. My generation and all that follow must take up their challenge. Serengeti and Mara can survive. But only if we really want them to.

Author's Note

While in the Mara I was fortunate in being able to base myself at Kichwa Tembo, a luxury tented camp lying just outside the northern boundary of the Mara Triangle. Here I could always re-supply with film and fuel, have my vehicle repaired, yet still be within easy reach of the animals. But to follow the migration into the Serengeti I needed to be self-sufficient.

Serengeti is a vast wilderness. It would have been impossible to have travelled to and from a lodge or campsite each morning and evening and still maintain contact with the animals and photograph them to best advantage. My vehicle would have to act as my home. A hundred litre fuel tank was built into the footwell behind the front seats which, combined with jerry cans, gave me a total fuel capacity of 240 litres – enough diesel to allow me to travel for 2,000 km without refuelling. Drums of diesel were stored with friends at the Research Centre and Ndutu Safari Lodge.

The back of the vehicle was modified to enable me to sleep comfortably inside the car. A large lockable wooden box, with twin compartments, was built along one side of the vehicle in which I could store essentials – tinned food, cooking utensils and vehicle spares. A table hinged to the front end of this box folded forward on to collapsible metal legs to create a two metre bed. Mattress and bedding were carried in a weatherproof canvas bedroll. Forty litres of drinking water were stored in the two plastic jerry cans which were filled from natural springs, or by friends living at Seronera. Two spare wheels rested behind a single back seat. A fluorescent light, which hooked up to the car battery, allowed me to write up my notes at night and to see what I was eating – though I hardly needed to be reminded – invariably corned beef. Friends who visited me for supper were not impressed. Camping gas fitted with a cooking head served as a stove to cook an evening meal and boil water for endless cups of coffee.

A heavy duty winch proved invaluable when no other help was available. It acquired most use in pulling other people from the Mara's black cotton soils, though I did spend an interesting morning digging a tyre-sized hole in the middle of the Serengeti plain so as to bury my spare wheel as an anchor to free myself from a concealed warthog burrow. Better by far to carry a boat anchor or deadman anchor for use in such treeless terrain.

I have always used Canon cameras, namely F1's, A1's and T90's. These and a comprehensive range of Canon lenses are stored in separate compartments sewn into a large canvas bag covering the double seat next to me. This prevents cameras from bouncing around and falling on to the floor when I hit a hole, and ensures that any one of four cameras – each fitted with a different lens – can be reached quickly so as to avoid unnecessarily missing a picture.

There is a trunk full of tripod heads, monopods, suction mounts, shoulder braces, pistol grips and quick release gadgets, sitting collecting dust in Nairobi. All have eventually been cast aside in place of a simple metal table that slots into pieces of piping securely attached to the inside of my car door. The table can be raised or lowered on its two metal legs by loosening the wing nuts that hold it firmly at the required height. Used in conjunction with canvas bean bags, it provides a rock-solid support for even the longest of telephotos. It also doubles as a very convenient coffee table.

My longest lens is a Canon 800 mm f5.6, which is incredibly sharp and can comfortably be used down to 1/30th second when braced firmly on the table, an impossibility with a conventional tripod head. This lens has endured the roughest handling. During the excitement of photographing a river crossing I knocked it to the ground – a drop of one and a half metres – yet it survived unscathed.

My favourite lens is a Canon 500 mm f4.5. The sharpness and colour balance of this particular lens is unbeatable. I also frequently use a 300 mm f2.8. A 1.4 converter provides an invaluable safeguard against suddenly losing the use of one of my longer lenses. When used with my 500 mm lens, it increases the focal length to 700 mm; likewise the 300 mm converts to a 420 mm. You only lose one stop of light and still retain the minimum focusing distance of the prime lens.

Other Canon lenses which I found particularly useful were: a 200 mm f2.8; a 35-105 mm f3.5 zoom which proved ideal for aerial photography; and a high quality 20-35 mm f3.5 wide angle zoom which paid dividends at river crossings, when one could fill the frame with wildebeest and zebra. I use skylight filters to protect the front element of my lenses from dust and scratches, and a polarising filter to enrich colours and enhance the contrast of cloud-filled skies.

Many people who come on safari take the majority of their photographs from the roof hatch. While there are times when one undoubtedly benefits from this aerial view – as when a group of lions are gathered around a kill – I prefer as low an angle as possible to try and recreate the animal's view point, allowing the subject to dominate the frame.

I use fast lenses rather than photographing on high speed film. All my pictures are taken on Kodachrome 64 slide film, which is fine grained and reproduces colours the way I see them. Another advantage is that the colour dyes do not shift with time – an essential requirement for anyone hoping to sell his material in the years ahead. All my film is processed in Switzerland.

The pen and ink sketches were drawn in Indian ink with Rotring graphic pens fitted with 0.1 and 0.2 nibs.

Anyone interested in finding out more about Mara/Serengeti or contributing to their future may like to contact these organisations:

Friends of the Masai Mara: c/o World Wildlife Fund
1255 23rd St. N.W.
Washington D.C. 20037
U.S.A.

Friends of the Serengeti: c/o Neil Silverman
301 East 79th St
New York, NY 10021
U.S.A.

Acknowledgements

I wish to express my gratitude to the governments of Kenya and Tanzania in granting me permission to live and work in their countries.

While many other parts of Africa have seen their wildlife resources dwindle, Masailand still harbours the finest game viewing areas in the world. Masai pastoralists roamed the Serengeti/Mara long before the area was opened up to tourism, peacefully co-existing with wild animals for centuries. Hopefully the transition from nomadic pastoralism to individual landownership will still enable the animals to survive.

In Tanzania, David Babu, Director of Tanzania National Parks, enabled me to stay in Serengeti while collecting material for this book. Without his considerable help my work would have been impossible. In his years as Chief Park Warden of Serengeti, David demanded the highest of standards, which today are upheld by Bernard Maregesi, who was extremely helpful and patient with any requests I made for information or assistance.

Dr Markus Borner and his wife Monica often helped remind me of the joys of good cooking and warm hospitality at their home in the Serengeti. Markus and many other scientists at the Research Centre gave freely of their time in sharing their knowledge of the Serengeti and in providing details of their work.

As film makers, Alan and Joan Root have done more than anyone else to capture the beauty and intrigue of the Serengeti, and to share it with people throughout the world. I knew of their films long before I came to Africa, and they continue to inspire me with the excellence of their work and the breadth of their knowledge. When I first arrived at Seronera they made me welcome at their house, and provided space for my equipment.

Aadje Geertsema and Margaret Kullander have transformed Ndutu Safari Lodge. Ndutu is undoubtedly *the* place to stay when the migration is on the plains. When I was following the wild dogs they provided me with space to store fuel and vehicle spares at the Safari Lodge. Aadje had also spent time living in her car, when studying servals in the Ngorongoro Crater, and at times she would appear at the wild dog den with all manner of edible presents. Their kindness was greatly appreciated.

There are few people associated with Serengeti who have not benefited in some way from Barbie Allen's kindness. She provides a life line to those living in Serengeti, helping where possible to keep them supplied with the essentials of life: shopping for spares, phoning on their behalf and keeping track of mail. Barbie is a true friend of all those fortunate enough to live in Serengeti.

Neil and Joyce Silverman made it possible for me to visit the Serengeti for six weeks in early 1986, while I was trying to initiate the migration book in Tanzania. Their periodic visits to the area while I was working on the plains replenished my depleted resources with a store of long-forgotten luxuries. Their generosity has helped finance a number of wildlife projects in Serengeti and Mara, and they continue to support the wild dog monitoring project. I greatly appreciate their friendship.

The Serengeti would not have been the same without Clare Fitzgibbon and John Fanshawe. It was John who originally directed me to the Naabi pack's den site, in 1987, thereby enabling me to maximise the chance of observing and photographing the puppies from the earliest age. And Clare patiently helped remind me how much our understanding of animal behaviour has evolved since I was at university. They cheerfully re-supplied me with the essentials of life from my store of goods at their house at the Research Centre, delivered letters and films and allowed me free rein in their collection of research literature. More than this, their warmth and enthusiasm heightened the joy of watching the Naabi pack.

In Kenya, I should like to thank the Narok Country Council, who administer the Masai Mara National Reserve, for allowing me to live and work in the Mara for the last eleven years. They have undoubtedly been the best years of my life.

Senior Warden Simon Ole Makallah and Warden James Sindiyo have been particularly generous with their time and advice and granted permission for aerial photography of the migration. Without Simon's help it would have been impossible to complete the various projects I have worked on over the years.

I still often cross paths with Joseph Rotich, head driver-guide for Jock Anderson of East African Wildlife Safaris. His unrivalled knowledge of the Mara, together with his great humour and dignity, have brought endless pleasure to those fortunate enough to accompany him on safari. Jock Anderson still generously tolerates my use of his Nairobi office, where Sarah Moller, Jacky Keith and Stephen Masika helped in countless ways to enable me to spend long periods away on safari.

Since 1981, Geoff and Jorie Kent have provided me with a base at Kichwa Tembo, their luxury tented camp in the northern Mara. Apart from the enormous comfort of living at such a facility, the comprehensive range of back-up services proved invaluable. In this respect I wish to acknowledge my thanks to everybody at the Abercrombie and Kent offices in Nairobi, Arusha, London and Chicago who responded with great efficiency when ever they were asked to help.

Peter and Alison Cadot always made me welcome in their house at Kichwa Tembo, regardless of when I arrived or how long I stayed. The new managers, Maurice and Monica Anami, have been equally hospitable and went out of their way to accommodate my every need. And Richard Chai and his maintenance staff at the vehicle workshop somehow managed to keep my Toyota Landcruiser roadworthy. The direct radio link between Kichwa Tembo and Signet Hotels in Nairobi proved invaluable. Richard Markham was of great support, and whenever I needed something urgently, Jimmy Musyimi at Signet responded with lightning efficiency – booking air tickets, posting packages, purchasing film or vehicle spares – all done quickly and cheerfully.

My thanks to David and Kim Penrose at Mara River Camp, and to the managers and staff of all the other camps in the Mara for their many acts of hospitality, and for providing details of the activities of various animals that I was trying to keep track of. Herbert Schaible of African Safari Club was particularly helpful in this respect, passing on all manner of interesting information on the leopards and wild dogs during my absence.

Boris Tismimiezky rescued me at a time of financial crisis, lent me his books and provided me with a place to recuperate from a bout of malaria. What more need I say?

Dr Holly Dublin, WWF Project ecologist in the Masai Mara, proved an invaluable source of up-to-date scientific information on the Mara and Serengeti. She patiently listened to a volume of questions, and where possible redirected my muddled thinking. Dr Hugh Lamprey of WWF, Nairobi, was equally generous with his time and advice.

By good fortune I was able to spend an afternoon in Nairobi in the company of Dr Richard Estes, who has long studied the behaviour of the wildebeest. He was able to provide me with some valuable insights into the world of these extraordinary creatures

Colonel T. S. Connor, DSO, KPM, continues to provide me with a home during my infrequent visits to Nairobi. His unfailing kindness over the years has been of great support to me. What luxury just to be able to sit somewhere quiet to sort through a mountain of slides, or complete a drawing – to find everything just as I had left it months before. It was through his generosity that I was able to accompany him on a safari to Tanzania in 1983, and rediscover the magic of the Serengeti.

I first met Gregory and Mary Beth Dimijian when they stayed at Kichwa Tembo for a month during the summer of 1987. They had visited Mara and Serengeti on a number of previous occasions and we soon became friends. Greg proved a fund of stimulating knowledge on all aspects of photography and biology, and on returning to the USA generously provided me with an update of literature relevant to my work. The intermittent deliveries of mail received at camp would have seemed far less exciting without letters from the Dimijians.

David Goodnow generously made it possible for a steady flow of Kodachrome 64 film to reach me in Mara and the Serengeti.

Mike Harries proved the perfect pilot to fly me over the migration when it arrived in the Mara. If the pictures are lousy it is certainly not his fault.

In Belgium, Rowena Johnson was a great source of encouragement and performed wonders on her word processor in typing up the manuscript. Rowena's sister Ronny Loxton, and her husband Robert, generously put us up at their home in Banbury and allowed us to monopolise their word processor – friends indeed.

In England, Pippa Millard always made me welcome at her home in London when I needed to visit my publisher.

Brian Jackman, wildlife correspondent with the *Sunday Times* in London, cast a professional eye over a rough version of the text. A few deft strokes of his pen made all the difference. We had worked together to produce *The Marsh Lions* – an experience which I value greatly. Brian has always been incredibly generous with his time and help since I started to write my own books.

Caroline Taggart, my editor at Elm Tree Books, worked marvels. She has been of tremendous support during the production of all my books – always encouraging and optimistic – never allowing me enough rope to hang myself, yet receptive to last minute changes where they added to text or design. Mike Shaw, my agent at Curtis Brown, was equally supportive.

My whole family has been an unfailing source of inspiration and encouragement to me in my work. My mother is one of those rare spirits who has proved indomitable under the toughest of circumstances, and seems to thrive on making difficulties seem little more than minor inconveniences. It is always a joy to spend time at home with her in England.

Many other people have helped me over the years – it would take another book to mention each and every one by name – you are remembered; my thanks to you all.

Bibliography

I have been fortunate in being able to draw on the wealth of literature detailing research carried out in the Serengeti/Mara over the last twenty-five years. Additional to these are the eloquent accounts of the Serengeti by Grzimek, Hayes, Matthiessen and Schaller. The most comprehensive account of the ecosystem is *Serengeti: Dynamics of an Ecosystem*, edited by Sinclair and Norton-Griffiths, and a paper by Hugh Lamprey, listed below. Though I have not cited individual references in the text I found the works of the following scientists and authors indispensable in attempting to tell the story of the migration – all of whom remain blameless for any inaccuracies, or for the inevitable simplifications that I have made in interpreting their work.

BELL, R. H. V.
1970 *The use of the herb layer by grazing ungulates in the Serengeti.* In 'Animal populations in relation to their food resources', ed. A. Watson, p. 111–123. Oxford: Blackwell

BOWKER, M.
1987 *Spying from on high.* International Wildlife (Sept./Oct.) pg. 22–23

BRAUN, H. M. H.
1973 *Primary production in the Serengeti: purpose, methods and some results of research.* Ann. Univ. d'Abidjan (E) 6: 171–88

CAPRA, F.
1982 *The turning point: science, society and the rising culture.* Simon and Schuster

CRONWRIGHT-SCHREINER, S. C.
1925 *The migratory springbucks of South Africa* London: Fisher Unwin

DE WIT, H.
1978 *Soils and grassland types of the Serengeti plain (Tanzania). Their distribution and interrelations.* D. Phil. thesis, Agricultural University of Wageningen

DUBLIN, H.
1984 *The Serengeti/Mara ecosystem.* Swara Vol. 7. No. 4

1986 *Decline of Mara woodlands: the role of fire and elephants.* University of British Columbia. D. Phil. thesis

ESTES, R. D.
1966 *Behaviour and life history of the wildebeest* (Connochaetes taurinus Burchell). Nature (London), 212: p. 999–1000

1969 *Territorial behaviour of the wildebeest* (Connochaetes taurinus Burchell). Z. Tierpsychol., 26, p. 284–370

1976 *The significance of breeding synchrony in the wildebeest.* E. Afr. Wild. J., 14, p. 135–152

GRZIMEK, B. AND GRZIMEK, M.
1960 *Serengeti shall not die.* London: Hamish Hamilton

HANBY, J. P., AND BYGOTT, J. D.
1979 *Population changes in lions and other predators.* In 'Serengeti: dynamics of an ecosystem,' eds. A. R. E. Sinclair and M. Norton-Griffiths, p. 249–262. Chicago: University of Chicago Press

1983 *Lions share: The story of a Serengeti pride.* London: Collins

HAY, R. L.
1970 *Pedogenic calcretes of the Serengeti Plain, Tanzania.* Abstr. with Program, Geol. Soc. Am 2: 572

1976 *Geology of the Olduvai Gorge.* Los Angeles: Univ. of Calif. Press

HAYES, H. T. P.
1977 *The last place on earth.* New York: Stein and Day

1981 *Three levels of time.* New York: Elsevier-Dutton

HOUSTON, D. C.
1979 *The adaptations of scavengers.* In 'Serengeti: dynamics of an ecosystem', eds. A. R. E. Sinclair and M. Norton-Griffiths, p. 263–286. Chicago: University of Chicago Press

JACKMAN, B. J., AND SCOTT, J. P.
1982 *The marsh lions.* London: Elm Tree Books

KINGDON, J.
1971–1982 *East African mammals: an atlas of evolution in Africa.* Vols. 1–7. New York: Academic Press

KREULEN, D. A.
1975 *Wildebeest habitat selection on the Serengeti plains, Tanzania, in relation to calcium and lactation: a preliminary report.* E. Afr. Wild. J. 13, p. 297–304

KRUUK, H.
1972 *The spotted hyena.* Chicago: Univ. of Chicago Press

LAMPREY, H. F.
1978 *The Serengeti region: a semi-arid grassland ecosystem in Africa.* In 'State of knowledge report on tropical grazing land ecosystems'. Geneva: UNESCO

LAMPREY, R. H.
1984 *The Masai: a society in transition.* In 'Animal Kingdom', Vol. 87, No. 3, p. 23–25. New York Zoological Society

MACCLINTOCK, D., AND MOCHI, U.
1984 *African images.* New York: Charles Scribner's Sons

MCNAUGHTON, S. J.
1979 *Grassland-herbivore dynamics.* In 'Serengeti: dynamics of an ecosystem', eds. A. R. E. Sinclair and M. Norton-Griffiths. p. 46–51. Chicago: Univ. of Chicago Press

1984 *Grasslands and grazers: a system in balance.* In 'Animal Kingdom', Vol. 87, No. 3, p. 36–38. New York Zoological Society.

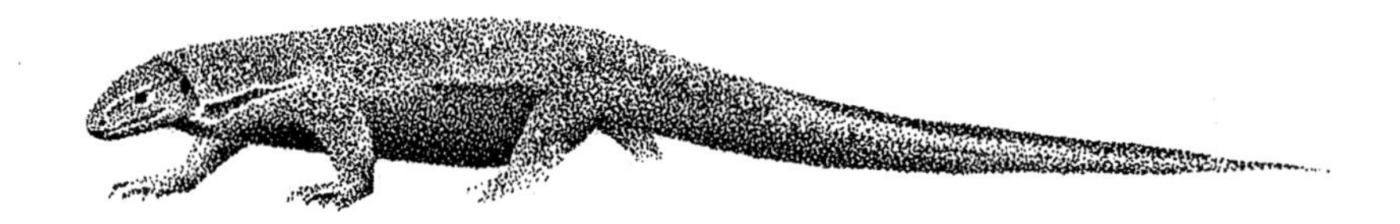

1985 *Ecology of a grazing ecosystem; The Serengeti*
Ecol. Monograph. 55. p 259–294

MADDOCK, L.
1979 *The 'migration' and grazing succession.*
In 'Serengeti: dynamics of an ecosystem' eds. A. R. E. Sinclair and M. Norton-Griffiths, p. 104–129.
Chicago: University of Chicago Press

MALPAS, R. AND PERKINS, S.
1986 *Toward a regional conservation strategy for the Serengeti.*
International Union for Conservation of Nature and Natural Resources

MATTHIESSEN, P.
1972 *The tree where man was born.*
London: Collins

MEAGHER, M.
1985 *Yellowstone's free-ranging bison.*
Naturalist: conservation through education.
Vol. 36 No. 3

1986 *Bison bison:*
Mammalian Species No. 266, p. 1–8.
The American Society of Mammalogists

MOSS, C.
1975 *Portraits in the wild.*
Chicago: Univ. of Chicago Press

MYERS, N.
1972 *The long African day.*
New York: Macmillan Pub. Co

OWENS, D. D., AND OWENS, M. J.
1985 *Cry of the Kalahari.*
London: Collins

PENNYCUICK. C. J.
1979 *Energy costs of locomotion and the concept of 'foraging radius'.*
In 'Serengeti: dynamics of an ecosystem', eds. A. R. E. Sinclair and M. Norton-Griffiths, p. 164–184.
Chicago: University of Chicago Press.

PENNYCUICK. L.
1975 *Movements of the migratory wildebeest population in the Serengeti area between 1960 and 1973.*
E. Afr. Wildl. J. 13. p. 65–87

READER, J., AND CROZE, H.
1977 *Pyramids of life.*
London: Collins

RICCIUTI, E. R.
1984 *Animal Kingdom.*
Vol. 87., No. 3., p. 12–47. New York Zoological Society

RUSSEL, F.
1984 *The hunting animal.*
London: Hutchinson & Co. (Pub.) Ltd

SCHALLER, G. B.
1972 *The Serengeti lion: a study of predator – prey relations.*
Chicago: Univ. of Chicago Press

1973 *Serengeti: a kingdom of predators.*
London: Collins

1974 *Golden shadows, flying hooves.*
London: Collins

SCOTT, J. P.
1985 *The leopard's tale.*
London: Elm Tree Books

SINCLAIR, A. R. E.
1977a. *The African buffalo: a study of resource limitation of populations.*
Chicago: Univ. of Chicago Press

1977b. *Lunar cycle and timing of mating season in Serengeti wildebeest.*
Nature (London) 267. p. 832–833

1984 *The wildebeest triangle.*
In 'Animal Kingdom' vol. 87, No. 3, p. 20–22.
New York Zoological Society.

1987 *Long term monitoring in the Serengeti-Mara: trends in wildebeest and gazelle populations.*
Report 5. Serengeti ecological monitoring programme

SINCLAIR, A. R. E., AND NORTON-GRIFFITHS, M., EDS
1979 *Serengeti: dynamics of an ecosystem.*
Univ. of Chicago Press

SMITHERS, R. H. N.
1983 *The mammals of the Southern African subregion.*
Univ. of Pretoria

TALBOT, L. M., AND TALBOT, M. H.
1963 *The wildebeest in Western Masailand, East Africa.*
Wildlife Monographs, no. 12. The Wildlife Society, Washington D.C.

TURNER, K.
1978 *Serengeti home.* London: George Allen and Unwin

TURNER, M.; JACKMAN, B. J., ED
1987 *My Serengeti years.*
London: Elm Tree Books

VESEY-FITZGERALD, D.
1973 *East African grasslands.*
Nairobi: East African Publishing House

WATSON, R. M.
1967 *The population ecology of the wildbeest (Connochaetes taurinus albojubatus Thomas) in the Serengeti.*
Cambridge University. D. Phil. thesis